Smart Climate Action Through Transfer of Development Rights

Rick Pruetz

Also by Rick Pruetz

Ecocity Snapshots: Learning from Europe's Greenest Places

The TDR Handbook: Designing and Implementing Transfer of Development Rights Programs (co-authored with Arthur C. Nelson and Doug Woodruff)

Lasting Value: Open Space Planning and Preservation Successes

Beyond Takings and Givings: Saving Natural Areas, Farmland and Historic Landmarks with Transfer of Development Rights and Density Transfer Charges

Saved by Development: Preserving Environmental Areas, Farmland and Historic Landmarks with Transfer of Development Rights

Putting Transfer of Development Rights to Work in California

DEDICATION

To Adrian, Jay, Erica, Gena, Jeromy, Josh, Kayla, Cate, Evie and Sienna

CONTENTS

Smart Climate Action

Land use is essential to climate action. In addition to renewable energy and electric vehicles, we need diverse, compact communities where people can meet their daily needs without a car. We need to preserve our farms, forests, and wetlands, as well as maximize their ability to sequester carbon. We need to secure the embedded energy in historic landmarks, safeguard our water, restore biodiversity, and adapt to the growing threat from wildfires, floods, and sea level rise.

Land use actions are also synergistic. A greenbelt may curb energy-wasting sprawl while additionally preserving a forest, which in turn helps protect water resources, restore ecosystems, and reduce human exposure to wildfires and floods. In order words, land use actions can generate considerable bang for the buck.

That said, land use actions are still expensive. However, there are cost-effective ways of redirecting growth away from inappropriate places to locations where development can be safely and efficiently accommodated. One of these ways is transferable development rights. TDR operates within a jurisdiction's land use regulations, allowing additional development potential in optimal locations when developers contribute to the preservation of places in need of protection, including farmland, forests, and vulnerable coastal zones. Unlike other preservation tools, TDR is powered by private sector profits rather than taxation, a feature that appeals to already-burdened taxpayers. In addition, jurisdictions themselves can buy and sell TDRs, transforming initial seed money into a perpetual revolving fund for preservation.

Even though TDR has been in existence for over five decades, only a small percentage of US jurisdictions use it. Some cities and counties are concerned about TDR's complexity (although many jurisdictions have pioneered relatively simple, user-friendly innovations). TDR also has a spotty track record (however, TDR programs can succeed by observing well-known factors). Furthermore, even though TDR is market driven, it nevertheless requires expenditures in the form of public involvement, public-private cooperation, and political will. However, the cost of using TDR for climate action remains much lower than the cost of inaction.

Climate Change and Land Use

Governmental action at the local, state and national levels is urgently needed to address climate change. In its 2018 report, the International Panel on Climate Change (IPCC) analyzed likely impacts assuming humans are able to limit global warming to 1.5 degrees centigrade above pre-industrial levels. The risks grow if we miss the 1.5-degree target and reach 2.0 degrees: drought, heavy precipitation events, ice sheet loss, sea level rise, flooding, coastal inundation, saltwater intrusion, ecosystem failure, species extinction, wildfire, biodiversity decline, ocean acidification/anoxia, coral depletion, poverty, hunger, vector-borne disease, and life-threatening heatwaves. To address this challenge, the IPCC calls on individuals, businesses, non-governmental organizations, and all levels of government to adopt climate action mitigations and adaptations (IPCC 2018). In its 2021 report, Climate Change 2021: The Physical Science Basis, the IPCC emphasizes that climate change is not in the future but here with us now. "It is unequivocal that human influence has warmed the atmosphere, ocean and land. Widespread and rapid changes in the atmosphere, ocean, cryosphere and biosphere have occurred" (IPCC 2021 p6).

Urban form, land use and food production practices are essential climate actions. Drawdown, a 2017 study that quantified the carbon-reducing capabilities of 100 strategies, found that one third of its total estimated greenhouse gas (GHG) reductions could be achieved by combining 25 strategies involving compact urban development, planet-friendly food/fiber/biomass production plus the conservation of forests and wetlands (Hawken 2017).

Another 2017 study similarly estimated that 37 percent of necessary, cost-effective CO2 mitigation can be provided by natural climate solutions: conservation, restoration, and/or improved land management strategies that promote carbon storage and/or eliminate GHG emissions from forests, wetlands, grasslands, and agriculture (Griscom et al. 2017). Similar conclusions were reached by the IPCC in its 2019 special report: Climate Change and Land (IPCC 2019).

TDR is a logical climate change mitigation strategy considering its ability to encourage the redirection of development potential into cities and away from the natural areas and farmland needed for carbon sequestration and local food production. Governments adopt TDR regulations offering property owners the potential to be compensated for voluntarily preserving these resource lands. Unlike most other land preservation programs, the money for this compensation does not come from public sources or tax deductions. Instead, developers provide this compensation in return for the ability to achieve more profitable forms of development in locations where higher density is suitable and desirable.

By preserving farmland, forests, greenbelts, and other places that mitigate GHG, TDR also implements adaptation in locations increasingly threatened by flooding, wildfire, sea level rise and other hazards exacerbated by climate change. For example, by combining projections of population growth/migration and sea level rise, some researchers estimate that more than one billion people in low elevation coastal zones worldwide could be impacted by 2060 (Neumann et al. 2015). Similarly, adaptation strategies are needed to reduce risk from wildfire, to protect biodiversity and address other challenges enumerated by the IPCC. An effective TDR program offers the owners of vulnerable land the option of reducing or eliminating their development potential by selling it to allow additional growth in urban locations that reduce risk and facilitate a low-carbon way of life. Not all landowners will participate immediately. But in the long term, the growing threat of wildfires and coastal storms is likely to convince more at-risk property owners to choose the TDR alternative.

Non-governmental organizations as well as national and state level agencies agree that TDR should be considered for climate change mitigation and adaptation. The Policy Guide on Planning & Climate Change from the American Planning Association (APA) lists TDR as an option for preserving natural ecosystems and farmland in order to sequester carbon and reduce vehicle miles traveled (VMT) by promoting local food production (APA 2011). The United States Environmental Protection Agency and the United States National Oceanic and Atmospheric Administration (NOAA) list TDR as a way of relocating development potential away from vulnerable coastal areas to inland locations (NOAA 2012; Titus 2011). The California Coastal Commission (CCC) and the South Florida Regional Planning Council (SFRPC) also include TDR as an implementation measure for managed retreat from rising sea levels (CCC 2018; SFRPC 2013). In its Adaptation Tool Kit, the Georgetown Climate Center noted that TDR can accomplish the same results as land and easement acquisition programs at less governmental expense (Grannis 2011). A study published by Columbia Law School also notes that TDR has the potential to shift development away from hazardous shores (Siders 2013).

TDR 101

TDR is a regulatory tool that uses private sector profits rather than tax revenue to implement various planning goals. TDR typically allows the potential for extra density or floor area in places where growth is planned and suitable when developers pay for the reduction or elimination of development potential in places less appropriate for development.

TDR programs often preserve farmland, historic landmarks and diverse environmental resources including forests, wetlands, surface water, ground water recharge areas, habitat, rare ecosystems, steep slopes and coastal areas. In addition to reducing human exposure to the rising threat of wildfires, floods, and sea level rise, this preserved land can have the effect of nurturing compact, energy-efficient communities that reduce greenhouse gas emissions.

The components of a TDR mechanism are established by a jurisdiction's zoning regulations. Typically, a TDR ordinance designates the area where it wants less or no development, called the sending area, and those places where extra development is wanted, called the receiving area. These areas can be designated by map and/or ordinance text. The ordinance creates a dual zoning framework allowing sending area property owners and receiving area developers the choice of using or declining TDR options.

When sending area property owners decline the TDR option, they can continue to use their land as allowed by the underlying zoning. When they choose to use the TDR option, sending area landowners in the most common TDR programs place a conservation easement on their property that permanently restricts or eliminates on-site development potential but allows those land uses that are consistent with the program's goals as spelled out in the easement. In many environmental programs, sending area owners may have the option of transferring title to a public agency or private, non-profit conservation organization. In return for these easements or title, property owners who choose to participate receive transferable development rights, or TDRs. In traditional programs, transactions use actual TDRs which are transferred from sellers to buyers using various legal instruments.

However, as described throughout this book, some TDR programs use simultaneous transfers of development potential in ways that do not require the issuance or exchange of actual TDRs. In other programs, actual TDRs are not needed because the sending area owners are compensated directly with money rather than being issued TDRs which the owners subsequently sell in order to receive compensation.

Developers also have a choice when developing receiving area property. They can decline the TDR option and build at or below a baseline density, building height, floor area, lot coverage, or whatever form of additional development potential the program allows to those who use TDR. Alternatively, developers can choose to exceed baseline and build to the maximum development potential allowed by TDR when they comply with the requirements established by the TDR ordinance. Usually, compliance requires acquisition of the legislated number of TDRs from sending site owners, a TDR bank, or some other intermediary like a private, non-profit conservancy.

Receiving area developers are motivated to choose TDR when the additional profit generated by exceeding baseline more than offsets the added expense of paying for TDRs and building to higher levels of development. As detailed in Success Factor 5 in Part II, successful programs adjust the TDR allocation ratio (the amount of sending area preservation needed per TDR) and or the allowance ratio (the amount of extra receiving area development potential granted per TDR) so that participation in the program is beneficial to both the sending area TDR seller and the receiving area TDR buyer.

Although the mechanism is logical, TDR requires an understanding of the local real estate market and motivations of various stakeholders. Consequently, a relatively low percentage of jurisdictions adopt TDR programs. Furthermore, many adopted TDR programs do not meet expectations. However, the TDR programs that have preserved the most land identify key features that are essential or at least helpful to crafting a successful program. These ten success factors are fully explained with examples in Part II of this book.

User Manual

To be blunt, this book argues that counties, cities, towns, and villages should seriously consider using TDR for climate action. Land use is on the front lines of GHG mitigation and climate change adaptation in ways ranging from the containment of sprawl and the evolution of compact, energy-efficient communities to the preservation of farms, forests, wetlands, and habitat, which have the potential to sequester carbon while also reducing vulnerability to wildfire, floods, and sea level rise. As a bonus, these land use strategies are synergistic and deliver multiple benefits. As noted above, the preservation of a forested hillside can reduce risk from fires and floods, protect biodiversity, safeguard water resources, and create a greenbelt surrounding a compact, multi-functional community where inhabitants move about mainly under their own, health-improving power.

TDR is one of several ways to achieve these multiple, synergistic benefits. But most others rely primarily or entirely on public funding, a feature that is not popular with taxpayers. Conversely, TDR is market driven and offers elected officials an opportunity to act on climate change without alienating a majority of their constituents.

Part I of this book, Climate Change Mitigation and Adaptation Using TDR, includes nine chapters with each chapter using one or two TDR programs to illustrate how communities use TDR to tackle the following aspects of climate change:

Chapter 1: Urban Form – Montgomery County, Maryland
Chapter 2: Forests – King County, Washington
Chapter 3: Farmland – Calvert County, Maryland and Boulder County, Colorado
Chapter 4: Wetlands – Miami-Dade County, Florida and Collier County, Florida
Chapter 5: Historic Landmarks – San Francisco, California and New York City, New York
Chapter 6: Adapt to Sea Level Rise – Ocean City, Maryland and Sarasota County, Florida
Chapter 7: Adapt to Wildfire – Pitkin County, Colorado and Los Angeles County, California

Chapter 8: Biodiversity: Palm Beach County, Florida and San Luis Obispo County, California
Chapter 9: Water – New Jersey Pinelands, New Jersey and Central Pine Barrens, New York

Note that many of these TDR programs were adopted before climate change was on the minds of most local government officials. Several were initially launched to protect a particular resource such as farmland, historic landmarks, or environmental areas. But the protection of these resources became a significant form of climate action over time even though not all of these jurisdictions explicitly acknowledge that fact. Furthermore, because land use actions are synergistic, an individual TDR program typically provides more than one form of climate action. For example, by preserving a majority of its 93,000-acre Agricultural Reserve, Montgomery County, Maryland, created a greenbelt that fosters compact, energy-efficient communities, preserves its countryside for the protection of watersheds, wildlife, and outdoor recreation, as well as secures a place where restorative agriculture can sequester carbon while providing a source of locally produced food. The chapters in Part I feature one or two TDR programs but mention several others. Profiles of these TDR programs can be found in alphabetical order in Part III of this book. Some of these profiles are extensive and others are short. Often a longer profile of each profile can be found under the TDR Update tab at the web site www.SmartPreservation.net.

Part II of this book, *Building a Successful TDR Program*, begins with the admission that TDR performance has been underwhelming in many communities. However, this book argues that underperformance is due to ineffective program components rather than a fundamental flaw in the TDR mechanism itself. These components are referred to as success factors as examined in What Makes Transfer of Development Rights Work? Success Factors from Research and Practice, a paper published in the *Journal of the American Planning Association* (Pruetz and Standridge 2009). These ten success factors become the ten subchapters of Part II:
1) Demand for Bonus Development
2) Optimal Receiving Areas
3) Sending Area Development Constraints
4) Few or No Alternatives to TDR
5) Market Incentives
6) Certainty of Ability to Use TDR
7) Strong Public Support
8) Program Simplicity
9) Promotion and Facilitation
10) TDR Banks

Part III of this book presents profiles of 282 TDR programs in the United States in alphabetical order. As mentioned above, some of these profiles are extensive and others are short. Often a longer version of shorter profiles can often be found under the TDR Update tab at the web site www.SmartPreservation.net.

References
APA. 2011. Policy Guide on Planning & Climate Change.
https://www.planning.org/policy/guides/adopted/climatechange.htm. Accessed 14 July 2019.

California Coastal Commission. 2018. Sea Level Rise Policy Guidance. https://www.coastal.ca.gov/climate/slrguidance.html. Accessed 14 July 2019.

Grannis, J. 2011. Adaptation Tool Kit: Sea Level Rise and Coastal Land Use. Georgetown Climate Center. https://www.georgetownclimate.org/files/report/Adaptation_Tool_Kit_SLR.pdf. Accessed 14 July 2019.

Griscom, B. et al. 2017. Natural Climate Solutions. Proceedings of the National Academy of Sciences of the United States. https://www.pnas.org/content/114/44/11645. Accessed 22 August 2019.

Hawken, Paul. (ed.) 2017. Drawdown: The Most Comprehensive Plan Ever Proposed to Reverse Global Warming. New York: Penguin Books.

IPCC. 2018. Global Warming of 1.5 Degrees Centigrade. https://report.ipcc.ch/sr15/pdf/sr15_spm_final.pdf. Accessed 16 July 2019.

IPCC. 2019. Climate Change and Land: IPCC Special Report on Climate Change, Desertification, Land Degradation, Sustainable Land Management, Food Security, and Greenhouse Gas Fluxes in Terrestrial Ecosystems. file:///C:/Users/Richard/Documents/IPCC%202019%20Land%20Use%20Study%204.-SPM_Approved_Microsite_FINAL.pdf. Accessed 23 August 2019.

IPCC. 2021. Climate Change 2021: The Physical Science Basis. Accessed 8-9-21 at https://www.ipcc.ch/report/ar6/wg1/downloads/report/IPCC_AR6_WGI_Full_Report.pdf.

Neumann, B et al. 2015. Future Coastal Population Growth and Exposure to Sea-Level Rise and Coastal Flooding – A Global Assessment. https://journals.plos.org/plosone/article?id=10.1371/journal.pone.0118571. Accessed 17 July 2019.

NOAA. 2012. Achieving Hazard-Resilient Coastal & Waterfront Smart Growth: Coastal and Waterfront Smart Growth and Hazard Mitigation Roundtable Report. https://coastalsmartgrowth.noaa.gov/pdf/hazard_resilience.pdf. Accessed 23 August 2019.

Pruetz, R. and Standridge, N. 2009. What Makes TDR Work? Success Factors from Research and Practice. Journal of the American Planning Association 75 (1): 78-87.

Siders, A. 2013. Managed Coastal Retreat: A Legal Handbook on Shifting Development Away from Vulnerable Areas. Columbia Law School. http://columbiaclimatelaw.com/files/2016/11/Siders-2013-10-Managed-Coastal-Retreat.pdf. Accessed 23 August 2019.

South Florida Regional Planning Council. 2013. Adaptation Action Areas: Policy Options for Adaptive Planning for Rising Sea Levels. http://southeastfloridaclimatecompact.org/wp-content/uploads/2014/09/final-report-aaa.pdf. Accessed 14 July 2019.

Titus, J, 2011. Rolling Easements. United States Environmental Protection Agency Climate Ready Estuaries Program. https://www.epa.gov/sites/production/files/documents/rollingeasementsprimer.pdf Accessed 14 July 2019.

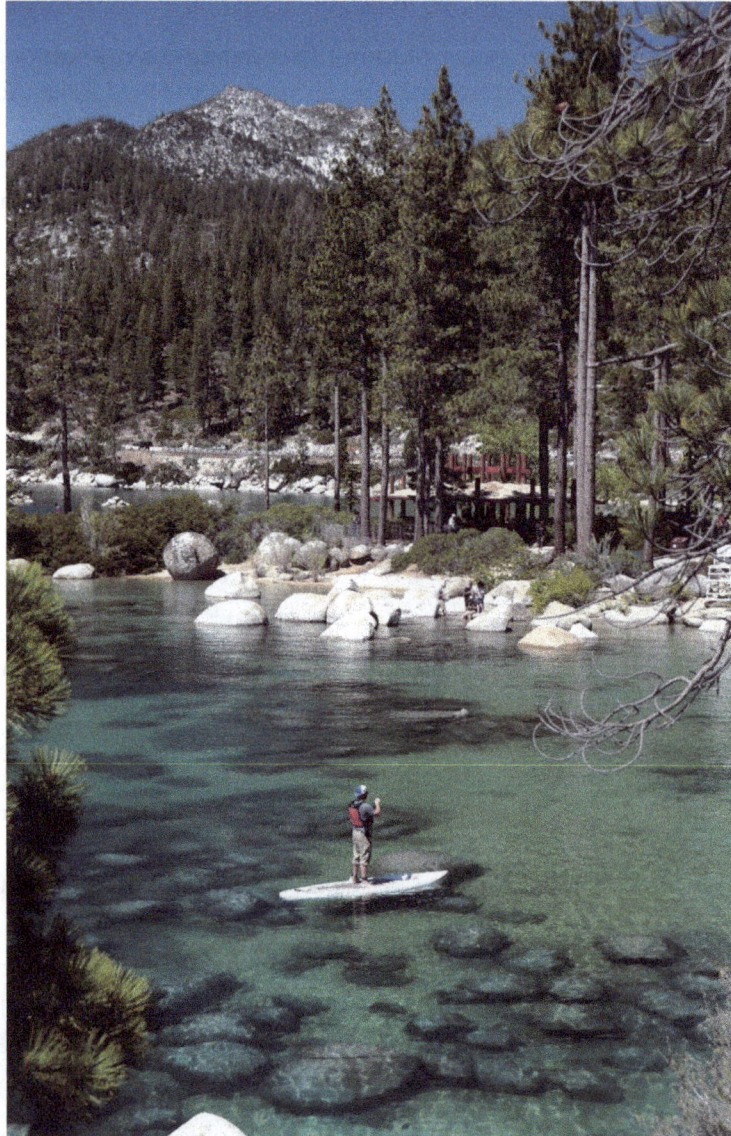

The TDR program managed by the Tahoe Regional Planning Agency, California/Nevada, aims to protect the water quality of Lake Tahoe.

PART I

Climate Change Mitigation and Adaptation Using TDR

The chapters in Part I demonstrate nine ways in which jurisdictions use TDR for climate action. Each chapter highlights one or two TDR programs and mentions other examples which can be found in the 282 TDR program profiles organized in alphabetical order in Part III.

CHAPTER 1

Urban Form – Montgomery County, Maryland

C ompact urban form, *by itself*, can cut resource and energy use *in half* concluded a report prepared by the International Resource Panel (IRP) for the United Nations Environment Programme. Compact urban growth is just the first of four levers of change recommended by the IRP in *The Weight of Cities: Resource Requirements for Future Urbanization*. The others are: livable, functionally and socially mixed neighborhoods; resource-efficient buildings as well as urban waste, water and energy systems; and sustainable human behaviors. Significantly, the IRP found that all four of these levers can each reduce energy and resource consumption at least in half. Plus these levers are also multiplicative: when they are employed in an integrated, mutually reinforcing manner, they can decrease resource use by 80 to 90 percent when compared with the resource consumption levels associated with current business-as-usual (BAU) practices (IRP 2018).

Over the last two centuries, most cities have generally become less dense, consequently decreasing the efficiency of embedded resources (those used in the construction of infrastructure and buildings) as well as the energy and other resources used in the operation of the built environment. Between 1800 and 2000, 28 of 30 world largest cities saw their densities drop by 1.5 percent per year. The price of urban inefficiency is clear in the United States, where the cost of sprawl has been pegged at $400 billion per year. This inefficiency can also be measured in energy waste and greenhouse gas (GHG) generation. The IRP report compares Atlanta with Barcelona, which have comparable populations and income levels. However, Atlanta is 27 times less dense than Barcelona and emits almost 11 times more GHG per person than Barcelona (IRP 2018).

Consistent with the time-tested principles of good planning, the IRP emphasizes that natural features should guide urban form. Greenspace can cool neighborhoods, manage rainwater naturally, offer sources of local food, provide safe pathways for non-motorized mobility, preserve ecosystems, and create greater contact between people and natural environments. To quote from the IRP report:

> *An interconnected system of natural spaces, ranging from a regional greenbelt to a pocket play park should, from a landscape ecology perspective, provide the main structuring elements of urban settlements. This principle reflects the importance of identifying natural systems and strategic landscape patterns, which protect valuable ecosystem services and biological hotspots, and designing the city around them (i.e. linking these systems if fragmented (IRP 2018, p 119).*

One of IRP's case studies comes from Copenhagen, Denmark, which created compact urban form by concentrating growth around major public transportation routes. This strategy left greenspace, known as green wedges, between these radial transit corridors. Realizing that the green wedges were vulnerable to suburban sprawl, Copenhagen and its neighboring jurisdictions filled them with parks, athletic fields, golf courses, forests, community gardens and landscape protection areas. As a result, average citizens were more likely to value these open spaces and Denmark officially adopted the green wedges in the National Planning Act of 2007.

Copenhagen has now become a world leader in eco-mobility and climate action. Between 2005 and 2014, Copenhagen's population grew by 15 percent and its economy rose by 18 percent while its carbon emissions fell by 31 percent. Currently, 62 percent of Copenhagen's work trips occurs on bikes and one third of suburban rail passengers use bicycles to solve the first/last mile problem. By 2025, Copenhagen wants eco-mobility to account for 75 percent of all trips and to make this city the world's first carbon-free capital (IRP 2018; Pruetz 2016).

Many cities restore as well as preserve greenspace in ways that generate diverse environmental benefits including compact urban form. To use another European example, Vitoria-Gasteiz, Spain, created a greenbelt around its compact urban area by restoring abandoned gravel pits, garbage dumps, the Zadorra River, and 247 acres of disturbed land into the internationally significant Salburua Wetlands. The city refers to this greenbelt as an eco-recreational corridor since it now offers 91 km of walking/bicycling trails. Significantly, the greenbelt also establishes the boundaries of Vitoria's urban area, where the city's process of re-densification succeeded in capturing 97 percent of the municipality's total growth between 2001 and 2010. As an indication that this strategy is working, even though the population of this city of 250,000 people has tripled since the 1960s, 81 percent of the population lives within 1,500 meters of the city center and over half of all trips here occur on foot (Pruetz 2016).

In the United States, Montgomery County, Maryland, used TDR to permanently preserve roughly two-thirds of the 93,000-acre greenbelt that now wraps around its development corridor, securing the compact urban form recommended by the IRP. When the county launched its TDR program in 1980, it aimed to curb sprawl, protect natural resources, and retain agriculture as well as agricultural land. But the greenspace preservation is inextricably tied to climate action, a fact that Montgomery County now explicitly states in its plans and policies.

As detailed by the Montgomery County profile in Part III of this book, this success story begins with a good plan, as is often the case with success stories. In 1964, Montgomery County adopted *On Wedges and Corridors*, a general plan aimed at concentrating growth within an urban spine that could be efficiently served by transportation and other public services, while saving the rest of the county as green wedges used for agriculture, watershed protection, wildlife habitat, outdoor recreation, and various other environmental benefits. The county tried to implement this plan by downzoning the wedges, an action that failed to keep the county from losing almost 12,000 acres of rural land to low density development in the first five years of the 1970s.

By 1980, Montgomery County diagnosed that the owners of rural land were suffering from "impermanence syndrome", a conviction that farming here was doomed because zoning alone was unable to keep sprawl from inevitably making farming difficult or impossible when located near encroaching residential subdivisions. The county identified a 93,000-acre Agricultural Reserve and downzoned it to minimum lot sizes of 25 acres. However, planners here understood that zoning alone is unable to stop sprawl in the long run. Using the TDR program described in Part III, the county motivated landowners in the Ag Reserve to record conservation easements permanently imposing the 25-acre minimum parcel size. In return, these participating sending site owners could sell one TDR for each five acres of land protected by these perpetual easements.

Receiving sites were located in the urban corridor, where developers could achieve a baseline level of development without using TDR. Alternatively, these developers could exceed baseline and achieve maximum density by buying and retiring TDRs.

Montgomery County observes most of the ten TDR success factors discussed in Part II of this book. The program established receiving area baselines that developers could profitably exceed despite the extra cost of buying the TDRs. The county established multiple receiving areas, often in single-family zones where a small increase in density could generate meaningful increases in net profit. The zoning imposed on the sending areas was adequate to motivate the owners of 52,052 acres to place their land under easement by 2013, with the $117 million in compensation paid by receiving site developers rather than taxpayers. The basic program components (for example, one TDR per five acres under easement in sending areas and one bonus single-family residential unit per TDR in receiving areas) created TDR prices that both sending area property owners and receiving area developers found attractive, creating an active TDR market. As a result, Montgomery County produced one of the nation's most successful programs despite the fact that it does not use a TDR bank.

Montgomery County launched its TDR program in 1980, before there was widespread recognition of the threat posed by climate change. But for over a decade, the county has identified TDR as one of its climate action strategies. The 2009 *Climate Protection Plan* recognized the success of the TDR program in reducing GHG emissions by surrounding its development corridor with a permanently preserved greenbelt. The plan explained that the urban core provides compact, mixed-use, livable places that "…invite(s) people to walk or bike to work, to shop and to participate in community life without a long commute by car. The Agricultural Reserve should continue to be protected for food production, recreation and carbon sequestration" (Montgomery County 2009, ES-9).

In 2017, Montgomery County declared a climate emergency and committed to reducing GHG emissions by 80 percent by 2027 and 100 percent by 2035. In June 2021, the county unveiled a Climate Action Plan aimed at implementing these ambitious goals using multiple strategies including ones that rely on the Agricultural Reserve largely preserved by TDR. Sequestration Action 4 - Regenerative Agriculture acknowledges the value of the Ag Reserve and recommends increasing its potential for improved carbon sequestration, biodiversity, water management, and natural ecosystems by increased regenerative agricultural practices including herbaceous field borders, reforestation in stream buffers, healthy soil practices, succession planting, cover crops, rotational grazing, no-till planting, and silvopasture systems that allow domesticated animals to graze in farms with forests. Action S-4 adds

16

that the county will need to incentivize regenerative agriculture using TDRs in the Ag Reserve. Furthermore, the plan aims to track TDRs in order to account for their effectiveness in reaching carbon sequestration goals (Montgomery County 2021).

As described elsewhere in Part I, Montgomery County is not the only US community that uses TDR to promote compact communities through greenspace preservation. Roughly half of the total land area of King County, Washington, is protected by various governmental agencies (Pruetz 2012) including 144,500 acres preserved by TDR for multiple benefits including climate action. Boulder County, Colorado uses TDR in interlocal agreements that helped create a greenbelt around the City of Boulder. Pitkin County, Colorado employs TDR to help keep development within its Urban Growth Boundary in an effort to reduce human exposure to wildfires and other hazards as well as curb sprawl and confine growth to areas that can be efficiently served by infrastructure. On a smaller scale, the receiving area of Old York Village in Chesterfield Township, Burlington County, New Jersey, is largely surrounded by farmland permanently preserved by TDR and other tools.

References

IRP (International Resource Panel). 2018. The Weight of Cities: Resource Requirements of Future Cities. Paris: United Nations Environment Programme.

Montgomery County. 2009. Climate Protection Plan. Accessed 6-27-21 at https://www.montgomerycountymd.gov/DEP/Resources/Files/downloads/outreach/sustainability/2009-moco-climate-protection-plan.pdf.

Montgomery County. 2021. Climate Action Plan. Accessed 6-27-21 at https://www.montgomerycountymd.gov/green/Resources/Files/climate/climate-action-plan.pdf.

Pruetz, Rick. 2012. Lasting Value: Open Space Planning and Preservation Successes. Chicago: Planners Press, American Planning Association.

Pruetz, R. 2016. Ecocity Snapshots: Learning from Europe's Greenest Places. Hermosa Beach: Arje Press.

CHAPTER 2

Forests – King County, Washington

orests not only sequester carbon but also maintain soil stability, promote water quantity/quality, provide shade, offer habitat to wildlife, and create opportunities for outdoor recreation and the study of nature. Many of the TDR programs profiled in Part III of this book use TDR to protect forests for their multiple benefits. King County, Washington, has been a leader in forest preservation accomplished by various levels of government including the King County TDR program which alone has protected 144,500 acres to date.

Our planet is home to 1.9 billion acres of temperate forests which sequester roughly 0.8 gigatons of carbon annually. By 2050, *Drawdown* estimates that yearly carbon sequestration could increase exponentially to as much as 22.6 gigatons by regenerating an additional 235 million acres of temperate forests around the world. Consequently, forest preservation and restoration is one of the leading strategies for mitigating climate change (Hawken 2017).

Mitigation strategies are often also adaptation strategies. Here, as elsewhere, forest preservation helps communities adapt to climate change by reducing the encroachment of urban development into places that are likely to be increasingly vulnerable to wildfires, floods, and other hazards as the planet continues to warm.

Controlling sprawl is a key goal of King County. Between 2011 and 2015, the county reduced VMT by maintaining compact urban form and efficient land use patterns. Specifically, 98 percent of all development here between 2011 and 2015 occurred within the Urban Growth Boundary (UGB). In 2015, King County adopted a *Strategic Climate Action Plan* (SCAP) that combines several preservation measures, including TDR, to curb sprawl, focus growth within energy-saving neighborhoods, sequester carbon in farms, forests, and other open space, and reduce the threat of climate-driven impacts like floods.

As detailed in the Part III profile, King County has been a leader in using inter-jurisdictional TDR transfers from rural areas into incorporated cities including Seattle, Bellevue, Issaquah, Sammamish and Normandy Park. These interjurisdictional transfers are governed by interlocal agreements that often specify where in the county the TDRs will come from, how many TDRs the receiving areas will accept, and the extent to which the county will incentivize the city to accept TDRs from land under county jurisdiction. In a 2001 interlocal agreement, Seattle agreed to accept TDRs from county farmland that allowed a 30 percent height bonus for buildings within the city's Denny Triangle, a downtown district in need of revitalization. In return, King County committed to apply $500,000 toward the installation of amenities in the Denny Triangle including green streets, pedestrian/bicycle improvements, transit facilities/incentives, open space, storm water management, public art, and street furniture.

In 2011, King County commissioned a study to estimate the GHG mitigation resulting from the transfer of the 70 TDRs from rural King County into downtown Seattle under the 2001 TDR agreement. The study calculated that the transfer of all 70 TDRs could eliminate 19,000 metric tons of GHG emissions over a 30-year time period (Williams-Derry & Cortes 2011).

The State of Washington has added motivation for jurisdictions in the Puget Sound Region to participate in a regional interjurisdictional program aimed at transferring development potential from rural farms and forests into cities. Specifically, cities that adopt TDR programs capable of accommodating their fair share of development potential now qualify to use tax increment financing to fund infrastructure, a tool that otherwise is unavailable in Washington. In 2013, Seattle and King County signed the first interlocal agreement to qualify under this regional TDR program. Per this agreement, Seattle creates a receiving area for 800 TDRs from sending sites under county jurisdiction and King County pledges to dedicate up to $15.7 million of additional property tax revenue to pay for open space and transportation improvements in Seattle's South Lake Union and Downtown districts.

The many reasons why King County's has successfully used TDR to protect 144,500 acres of resource lands can be found in the Part III profile. The discussion in this chapter focuses on the county's use of TDR for climate action. King County included TDR in its SCAP as one of many implementation tools aimed at achieving its goal of reducing GHG emissions 50 percent by 2030 and 80 percent by 2050 compared with a 2007 baseline. The 2015 SCAP cites the county's TDR program for its multiple benefits of conserving forests, preserving farmland, and curbing urban sprawl, actions that mitigate GHG emissions and sequester carbon. In its report card on the 2015 SCAP, one of the top ten accomplishments was the Land Conservation Initiative (LCI), King County's accelerated protection of the best and last remaining open spaces, farmlands, forests, parks, and trails. TDR is recognized in the highlights of this Land Conservation Initiative. For example, interjurisdictional transfers between King County and the City of Sammamish preserved some of the envisioned greenbelt while protecting habitat for salmon and other critical species.

In May 2021, King County approved its 2020 SCAP. It again recognized TDR as a key tool for preserving forests and recommitted the county to its Land Conservation Initiative (LCI) which aims to preserve 65,000 acres of high-conservation-value, currently unprotected lands within 30 years. The LCI calls for accelerating the preservation of farms and forests by redoubling efforts on existing market-based conservation programs like transfer of development rights.

The LCI's emphasis on TDR is not surprising considering that TDR was involved in the protection of at least 45 percent of the 200,000 acres of forest land protected by King County since 1970. In the largest single transaction, King County used $22 million of Conservation Futures tax revenue to buy 990 TDRs representing a conservation easement placed on the 90,000-acre Snoqualmie Forest. The acquisition of TDRs, rather than a traditional purchase of development rights approach, allowed the county TDR bank to hold and sell these TDRs, thereby transforming what would otherwise be a one-time use of money into an ongoing revolving fund for preservation.

As shown in the Part III profiles, many jurisdictions use TDR to preserve forests, often in conjunction with other valuable resource land. For example, TDR is one of several tools used to protect forests as well as specialty farms, wetlands, and water resources in the New Jersey Pinelands, a one-million-acre region in the southeastern quadrant of New Jersey. In 2007, New Jersey adopted a Global Warming Response Act aimed at offsetting 46 percent of the state's GHG emissions using forests and other natural carbon sinks. The New Jersey Pinelands region is home to some of the largest unbroken tracts of forest lands in the eastern United States. The 1980 plan for the New Jersey Pinelands required 53 municipalities and seven counties to conform their zoning to achieve resource protection goals and implement a TDR program. Receiving areas in 22 municipalities were formed to accommodate TDRs from agricultural and forested sending sites throughout the region, making this the largest interjurisdictional TDR program in the US. As of June 31, 2020, this program had preserved 55,391 acres in these sending areas.

Other TDR programs with an explicit forest preservation goal are profiled in Part III: Bellevue, Washington; Calvert County, Maryland; Central Pine Barrens, New York; Island County, Washington; Jericho, Vermont; Kittitas County, Washington; Lee, New Hampshire; Miami-Dade County, Florida; Park City, Utah; Scarborough, Maine; and Snohomish, Washington.

References

Hawken, P. 2017. Drawdown: The Most Comprehensive Plan Ever Proposed to Reverse Global Warming. New York: Penguin Books.

Williams-Derry, Clarke & Erik Cortes. 2011. Transfer of Development Rights: A Tool for Reducing Climate-Warming Emissions – Estimates for King County, Washington. Seattle: Sightline Institute.

CHAPTER 3

Farmland – Calvert County, Maryland and Boulder County, Colorado

Farmland is essential to climate action. As a driver of climate change, the generation of food currently emits almost as much GHG worldwide as the generation of electricity. Conversely, farmland can also sequester carbon using strategies like forest restoration, restorative agriculture, and perennial cropping. The second part of this chapter addresses agricultural practices. The first part deals with farmland preservation because in order to get the benefit of improved agricultural practices we must have farmland where we can apply these carbon-sequestering improvements.

Farmland Preservation

According to the American Farmland Trust (AFT), 2,000 acres of US farmland are converted to development *every day*, totaling 11 million acres between 2001 and 2016 (American Farmland Trust 2020). AFT's ubiquitous bumper sticker reads: "No Farms, No Food." Although it might not fit on a bumper, an equally true sticker could read: "No Farms, No Carbon Sequestration on Farmland."

The loss of farmland to sprawl in the US has been recognized as a serious problem for decades. In response, permanent farmland preservation has occurred, mostly through conservation easements and purchase of development rights (PDR). PDR uses public funds to buy permanent conservation easements. PDR is less complicated and less contentious than TDR because it does not require the relocation of development rights to receiving sites. Of course, its major drawback is that PDR typically relies on tax revenues, and, in some communities, it can be difficult to convince people that they should be taxed in order to preserve farmland. Conversely, TDR has more moving parts than PDR, as explored in Part II of this book. But, unlike PDR, TDR is funded by private sector developers rather than public revenue, a distinct advantage in those times and places where taxpayers feel inordinately burdened. Consequently, many communities choose to permanently save farmland using TDR either exclusively or in conjunction with PDR and other preservation mechanisms. Of the 247 programs listed in *The TDR Handbook*, over half use TDR primarily or partially for farmland preservation (Nelson, Pruetz & Woodruff 2012).

As described above, Montgomery County, Maryland, has one of the most successful TDR farmland preservation programs in the US, with over 52,000 acres preserved by TDR alone. Over 30,000 of the 55,391 acres preserved to date by the New Jersey Pinelands TDR program are in Agricultural Production Areas. However, since Part I uses Montgomery County as an example of greenbelt preservation and the New Jersey Pinelands as an example of water resource protection, this chapter offers Calvert County, Maryland for its example because this county 35 miles southeast of Washington, DC has preserved 23,000 acres of farmland with TDR.

Calvert County adopted its first TDR program in 1978, making it the first TDR program in Maryland. It was primarily designed to preserve farms and forests, but Calvert County uses TDR and land preservation in general as a climate action tool that mitigates the GHG emissions caused by sprawl, protects local food sources, and helps property owners adapt to wildfire, flood, storm surge, sea level rise and other hazards exacerbated by climate change.

Properties in three zoning districts that account for most of the land area in Calvert County can become either sending or receiving sites. Owners who wish to become sending sites apply for a designation of Agricultural Preservation District (APD). If approved as APD, the county certifies the number of TDRs allocated to the property. The APD status is in effect for at least five years. During that time, the site must be actively farmed or forested. If no TDRs are sold after the five-year period, the owner may remove the APD designation. However, after the first TDR is sold, a permanent easement must be placed on the entire property. The number of residential lots allowed by the permanent easement varies depending on the size of the APD as detailed in the Calvert County profile in Part III.

Alternatively, owners of land in these three zoning districts can apply to become receiving sites and increase allowable density at the ratio of five TDRs per bonus residential lot. The baseline density and bonus density varies depending on the property's zoning district and proximity to a town center. For example, a parcel zoned Residential within a one-mile radius of a Town Center has a baseline of one unit per four acres and can achieve a maximum density of four lots or units per acre, which represents a 16-fold increase by using TDR.

Calvert County has adopted major revisions to its program since 1978. In 1993 and 2003, the county downzoned baselines in some zoning districts from an original density of one unit per five acres to the current baseline of one unit per 20 acres. These downzonings increased the motivation of the owners of productive resource lands to become sending sites rather than develop to the relatively low baseline density allowed on site. The downzonings likewise encouraged owners of property that qualified to become receiving sites to use the TDR option in order to exceed the lower baselines allowed after these downzonings. Studies of this program attribute much of the success of the Calvert County program to these downzonings (McConnell, Kopits & Walls 2003; Walls 2012).

Calvert County is an active participant in the TDR market. In 1993, the county started the Purchase and Retirement (PAR) program, which buys and retires TDRs. In a second program adopted in 2001, Leverage and Retire (LAR), farmers preserve their land and are reimbursed by the County over time.

For example, a landowner might receive tax-free interest payments over 15 years followed by a payment of principal at the end of 15 years. The County benefits by being able to protect more land with limited near-term expenditure. However, landowners also benefit by deferring income into years when they plan to be retired and earning less income from other sources. The PAR and LAR programs contributed to price stability over time. But observers also note that the purchase price paid by the county can inhibit private market sales when sending site owners decline to sell to developers because they are hoping to be able to sell their TDRs to the county for a higher price.

By 2019, the TDR program in Calvert County had preserved over 23,000 acres, which is over half of the county's 40,000-acre goal. Because land in three zoning districts can become receiving sites as well as sending sites, the program has been criticized for contributing to the fragmentation of agricultural areas. However, the density allowed via TDR in the two largest zones is one unit per ten acres, a density that program supporters see as a reasonable tradeoff for accomplishing the amount of permanent preservation achieved in Calvert County.

Cultivating Better Agricultural Practices

As mentioned above, preservation must be accompanied by improved agricultural practices in order to maximize farmland's potential to mitigate and sequester carbon. According to a Project Drawdown report, *Farming Our Way Out of the Climate Crisis* (Project Drawdown 2020), six percent of current worldwide GHG emissions comes from methane caused by farm animals and rice cultivation, four percent from nitrous oxide generated by fertilizer and manure, and another five percent from other agriculture-related activities such as fuel consumed in producing agricultural chemicals, operating machinery, and transporting farm supplies and products.

In addition, destructive agricultural practices, including deforestation and land clearance for crops and pastures, destroy habitat. Agriculture is also responsible for 85 percent of water use on our planet. Overuse of fertilizers and manure pollutes streams, lakes and oceans around the globe as well as exacerbating climate change by sending excess nitrogen into the atmosphere.

In addition to mitigating GHG emissions, improved agricultural practices would increase carbon sequestration in biomass and soil. The strategies include forest restoration, regenerative annual cropping to rebuild soil, and increased adoption of perennial crops using agroforestry, intercropping trees with annual crops, and farming approaches that use multiple layers such as shade coffee. In addition to sequestering carbon, regenerative agriculture improves water retention, reduces erosion, and improves agricultural productivity, ideally freeing more land for agroforestry and perennial cropping, the farming practices that generally produce higher sequestration rates.

Once sequestered, carbon can again be lost if carbon-friendly practices are abandoned or if carbon is released by drought, fire, and floods – events that will become increasingly common with climate change. Consequently, Project Drawdown warns that permanent management is critical, particularly if carbon credits are being sold based on sequestration projections.

As reported above, Montgomery County, Maryland, wants to become a leader in improved agricultural practices as well as farmland preservation. Montgomery County pledges to reduce GHG emissions by 80 percent by 2027 and 100 percent by 2035. The county's Climate Action Plan, adopted in June 2021, aims to implement these ambitious goals partly using strategies that rely on the Agricultural Reserve, which was preserved largely by TDR. Sequestration Action 4 - Regenerative Agriculture acknowledges the value of the Ag Reserve and recommends increasing the ability of the Agricultural Reserve to sequester carbon, protect biodiversity, and provide water management by improving regenerative agricultural practices including herbaceous field borders, reforestation in stream buffers, healthy soil practices, succession planting, cover crops, rotational grazing, no-till planting, and silvopasture systems that allow domesticated animals to graze in farms with forests. Action S-4 adds that the county will need to incentivize regenerative agriculture using TDRs in the Ag Reserve. Furthermore, the plan aims to track TDRs in order to account for their effectiveness in reaching carbon sequestration goals (Montgomery County 2021).

Boulder County, Colorado, a recognized leader in open space conservation, is also exploring ways to boost carbon sequestration on farms and rangeland. In addition to 137,308 acres held by the federal government, Boulder County itself protects 105,386 acres by ownership, lease and easement (Pruetz 2012). As detailed in Part III of this book, TDR is one of several preservation strategies used by Boulder County.

In 1989, Boulder County added TDR to its clustering procedure in order to motivate the preservation of land identified in Boulder County's Comprehensive Plan as having agricultural, natural, cultural, environmental, scenic, community buffer or open space significance. The receiving sites cannot be located within areas likely to be annexed by a municipality without the approval of that jurisdiction.

Boulder County then added a second TDR tool called Transferred Development Rights Planned Unit Development (TDR/PUD) with the purpose of preserving agriculture, rural character, open space, scenic vistas, natural features, and environmental resources. The TDR/PUD process can work inter-jurisdictionally as well as intra-jurisdictionally. Boulder County uses intergovernmental agreements (IGA) to facilitate the transfer of development potential from sending sites within unincorporated Boulder County to receiving sites within incorporated cities. Boulder County has signed IGAs with eight municipalities and the unincorporated town of Niwot. The sending areas identified in these IGAs are properties under county jurisdiction near the participating communities that these cities have a particular interest in preserving. For example, the TDR provisions in the IGA between Boulder County and the City of Boulder was one of several implementation tools responsible for the outstanding greenbelt that now surrounds the city.

By 2005, over 85 percent of county land was either within incorporated communities or already preserved by public ownership or conservation easements resulting from federal and state programs as well as the many preservation options launched by Boulder County and its municipalities (Fogg 2005).

In 2008, Boulder County adopted a new program which requires developers to acquire transferable development credits (TDCs) to exceed a floor area baseline of 6,000 square feet within individual dwelling units. This 6,000-square foot baseline includes garages, basements and residential accessory structures. Each TDC allows an additional 500 square feet of floor area until reaching 1,500 bonus square feet of floor area. Above that 1,500 square foot increment, two TDCs are needed for every 500 square feet of additional floor area.

Sending sites for the TDC program are legal building lots with legal access. The number of TDCs available to a sending site increases as the sending site dwelling is limited to increasingly smaller floor area, with up to ten TDCs granted to owners that preclude any dwelling on their land.

In 2018, the Boulder County Environmental Sustainability Plan established a target of lowering GHG emissions 45 percent by 2030 (baseline 2005) using multiple strategies including the preservation of farmland and increased sequestration of carbon through regenerative agricultural practices. In 2019, Boulder County issued an update, reporting that the county had initiated a pilot carbon sequestration project to gauge the feasibility of increasing carbon sequestration within the soils of agricultural fields, grasslands/rangelands, forests and urban developments. The county sees carbon sequestration as a way to implement various climate goals including GHG emission reduction, better soil health, increased crop yields, additional water retention, and restoration of native plant and animal species. The county is working with experts at Colorado State University on a demonstration project using soil remediation to boost carbon sequestration on degraded rangeland. Additionally, the county supports producer-based carbon farming trials aimed at incorporating carbon farming strategies on farmland owned by Boulder County and developing carbon-farming systems that can be used on irrigated cropland and pastures.

Other Farmland Preservation TDR Programs that have achieved notable results are profiled in Part III including: Adams County, Colorado; Blue Earth County, Minnesota; Charles County, Maryland; Chesterfield Township, Burlington County, New Jersey; Douglas County, Nevada; Howard County, Maryland; King County, Washington; Lumberton Township, Burlington County, New Jersey; Manheim Township, Lancaster County, Pennsylvania; New Castle County, Delaware; New Jersey Pinelands; Payette County, Idaho; Pierce County, Washington; Puget Sound Region, Washington; Queen Anne's County, Maryland; Redmond, Washington; Rice County, Minnesota; San Luis Obispo County, California; St. Mary's County, Maryland; and Warwick Township, Lancaster County, Pennsylvania.

References

American Farmland Trust. 2020. Farms Under Threat: The State of the States. Accessed 6-29-21 at https://s30428.pcdn.co/wp-content/uploads/sites/2/2020/09/AFT_FUT_StateoftheStates_rev.pdf.

Fogg, P. 2005. Correspondence with author: February 9 2005.

McConnell, Virginia, Elizabeth Kopits, and Margaret Walls. 2003. How Well Can Markets for Development Rights Work: Evaluating a Farmland Preservation Program. Washington, D.C.: Resources for the Future.

Montgomery County. 2021. Climate Action Plan. Accessed 6-27-21 at https://www.montgomerycountymd.gov/green/Resources/Files/climate/climate-action-plan.pdf.

Nelson, C., R. Pruetz & D. Woodruff. 2012. The TDR Handbook: Designing and Implementing Transfer of Development Rights Programs. Washington, D.C.: Island Press.

Project Drawdown. 2020. Farming Our Way Out of the Climate Crisis. Accessed 3-15-21 at DrawdownPrimer_FoodAgLandUse_Dec2020_01c.pdf.

Pruetz, R. 2012. Lasting Value: Open Space and Preservation Successes. Chicago: Planners Press.

Walls, Margaret. 2012. Markets for Development Rights: Lessons Learned from Three Decades of a TDR Program. Washington, D.C.: Resources for the Future.

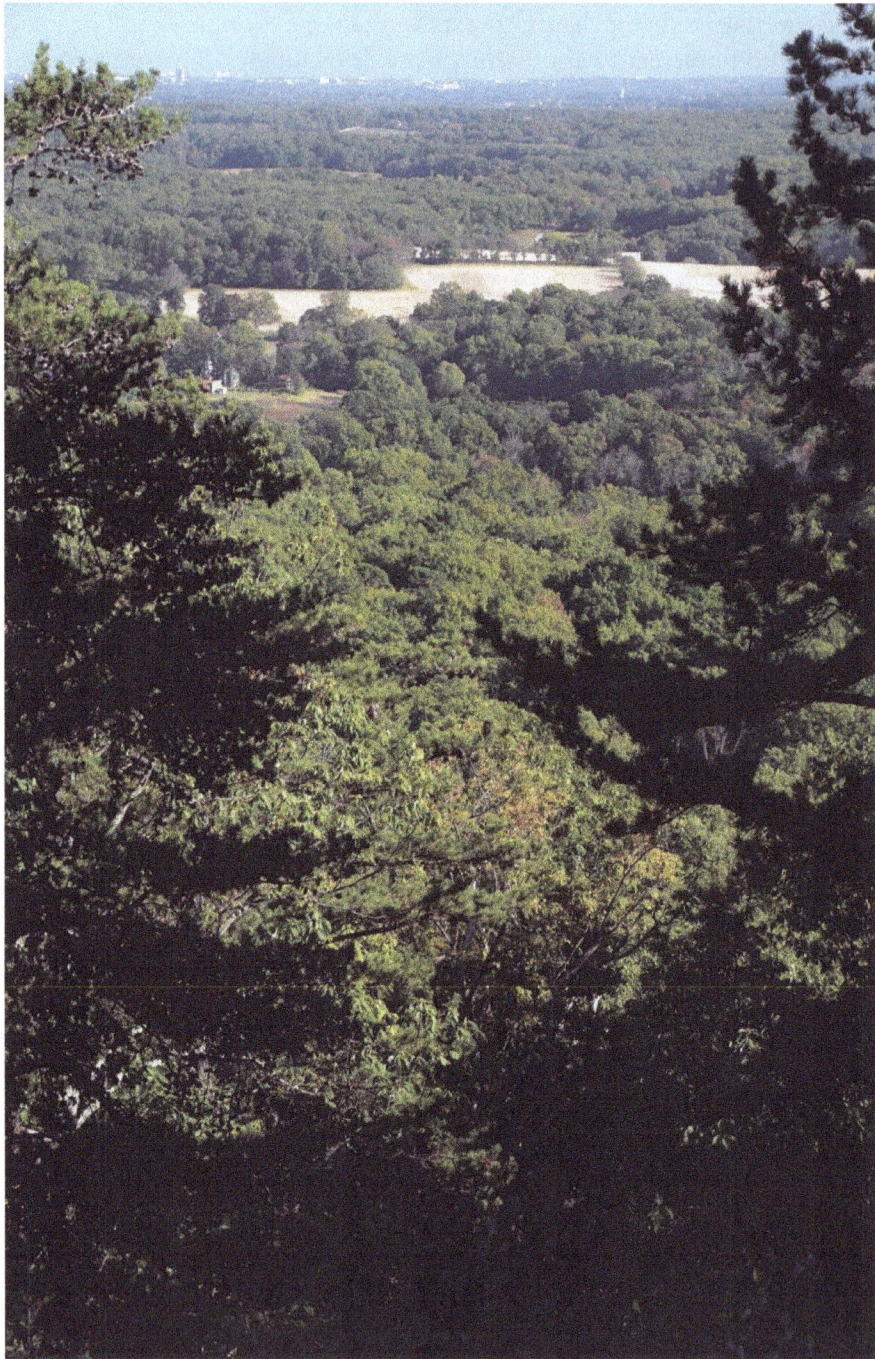

Montgomery County, Maryland has protected its 93,000-acre Agricultural Reserve mostly with TDR.

CHAPTER 4

Wetlands – Miami-Dade County, Florida and Collier County, Florida

Approximately one fifth of the planet's carbon is held by wetlands. Over extended time periods, coastal wetlands are capable of sequestering five times more carbon than tropical forests. However, human activity has already destroyed or degraded over one third of the world's mangrove forests, consequently releasing an alarming amount of long-sequestered carbon (Hawken 2017). The largest mangrove forests in the continental United States are located in the Florida Everglades. According to the National Science Foundation, a reliable supply of fresh water must be maintained in order to maintain the health of the Everglades' carbon-storing mangroves (National Science Foundation 2016).

Miami-Dade County, Florida

Miami-Dade County, Florida has been using TDR as part of its East Everglades Ordinance to safeguard the natural flow of water to Everglades National Park as well as protect the Biscayne Aquifer, the sole source of water for drinking and irrigation in Dade County as well as the Florida Keys.

Miami-Dade County, Florida, stretches from the dense coastal cities of Miami and its suburbs into the wetlands and mangrove forests of the Florida Everglades, which occupy the western and southern half of the county. As development threatened irreparable damage, Everglades National Park was established in 1934 and is now the third largest national park in the contiguous Unites States. The park is home to hundreds of bird, fish, reptile, and mammal species, including dozens of threatened or endangered species like the Florida panther and West Indian manatee, which largely explains the park's listing as a Biosphere Reserve, World Heritage Site, and Ramsar Wetland of International Importance.

In 1981, Miami-Dade County adopted its East Everglades Ordinance, which designated a 242-square mile area contiguous with Everglades National Park as an Area of Critical Environmental Concern in order to protect water quality, flood storage capacity, biodiversity, and the economic vitality of the county and its municipalities. This ordinance includes TDR provisions, which in Miami-Dade County are called severable use rights, or SURs.

The program has three sending areas encompassing roughly 45,200 acres. SUR allocations differ between these three sending areas. Owners of sending area land are limited to building at a maximum density of one unit per 40 acres although with stringent development requirements, including provisions to prevent disruption in the natural flow of surface water. However, when these owners protect their land by permanent easement, they can sever and sell the foregone development potential at the ratio of one SUR per five acres, one SUR per 12 acres, or one SUR per 40 acres depending on the location of the sending site within three management areas. Owners of legal lots smaller than these minimums can sell one SUR per lot if they registered within one year of program adoption. Lands traditionally submerged at least three months per year are considered undevelopable and cannot sell SURs.

Receiving areas are properties with residential, commercial, and industrial zoning designations located within Miami-Dade County's Urban Development Boundary (UDB). All Miami-Dade County zones designated for urban development can receive SURs with the exception of the agricultural, environmental, recreation, and open space zones. In four residential zoning districts, developers can use SURs to reduce minimum lot size and minimum frontage. In four additional residential zoning districts, SURs can be used to reduce maximum coverage as well as minimum lot size and frontage. In three more residential zones, SURs can increase maximum height limits as well as density, floor area ratio, and coverage. In two additional residential districts, SURs can increase maximum height, density and floor area ratio but not coverage. In the PAD, ECPAD, and REDPAD districts, SURs can increase density by 20 percent. In the Core or Center Sub Districts of Community Urban Center zoning districts with certain designations, baseline density can be increased by up to eight units at the rate of two units per SUR. In seven commercial districts, baseline floor area ratio (FAR) can be exceeded at the ratio of 0.015 FAR per SUR in seven commercial districts and by 0.010 FAR in the OPD district.

Importantly, receiving site developers can use SURs at receiving sites as a matter of right, which can reduce developer concerns about the delays, changes, and cost increases that sometimes result from discretionary approval. However, the receiving areas are subject to annexation by incorporated cities, thereby reducing the demand potential within the county. In response, as of 2015, Dade County allows SURs to be transferred inter-jurisdictionally to incorporated cities that agree to participate. As detailed in the Miami-Dade County profile in Part III, a 2017 report suggests that it may be necessary to require or encourage incorporated cities to create receiving areas for the interjurisdictional transfer of SURs (Miami-Dade County 2017).

Collier County, Florida

Collier County, Florida surrounds the City of Naples in the southwestern corner of Florida and includes portions of the Florida Everglades as well as numerous islands along its Gulf of Mexico shoreline. After decades of environmental damage caused by clear-cutting and ill-conceived real estate deals, Collier County has become increasingly environment friendly. Today, federal, state, private non-profit, and county efforts have preserved roughly 80 percent of Collier County in parks and preserves that include the Audubon Society's Corkscrew Swamp Sanctuary, Fakahatchee Strand Preserve, the Florida Panther National Wildlife Refuge, and Big Cypress National Preserve. The county's preservation efforts have included three different TDR programs.

Special Treatment Overlay - In 1974, Collier County created the Special Treatment (ST) Overlay zoning designation to protect coastal areas, wetlands, habitat, and other places of ecological significance. The ST overlay instituted strict environmental regulations and required approval of a special permit for all new development. As an alternative to on-site development, property owners were encouraged to preserve sensitive land using TDR.

The number of transferrable development rights is based on the density allowed by the underlying zoning of a sending site. When TDRS are transferred between urban areas, they can be transferred into all residential zoning districts and the residential components of planned unit developments. The maximum density allowed by the zoning of the receiving site serves as baseline and the bonus density is ten percent of baseline in the lower density residential zones (up to RMF-12) and five percent in the RMF-16, RT and PUD districts.

Rural Lands Stewardship (RLS) – The 300-square mile area surrounding the City of Immokalee, 30 miles northeast of Naples, is zoned for agriculture but contains many environmentally-significant resources including wetlands, flow-ways, water retention areas, aquifer recharge zones, and critical habitat. To head off protracted litigation, the major landowners came to an agreement with environmentalists and government officials in 1999 that produced a RLS program aimed at preserving the most sensitive areas and transferring the unused development potential to a new town located on less-sensitive land.

The program has a 195,000-acre planning area with the majority of the land under a few large ownerships. The agricultural zoning allows one on-site dwelling per five acres. The sending sites are proposed by the property owners and the stewardship credit allocation is calculated by a two-phase process. First, the land itself is assigned preservation value according to the priority status of its environmental resources, meaning habitat, flow-ways, or aquifer recharge zones. In the second step, the applicable property owners choose the extent to which they want to restrict the use of their property. Owners can decide to only prohibit residential development or also restrict other activities such as mining, recreation, cropland, agricultural support, and pasturing. The amount of credits available to a property grows as the owner chooses to apply greater restrictions.

The new town of Ave Maria is the receiving area for this RLS program. Here, one acre of receiving area land can be developed to the maximum extent allowed under the Ave Maria master plan when a developer retires eight credits. With this approach, the RLS program avoids the problem inherent in many TDR programs in which developers must retire an increasing number of TDRs to reach a maximum density that may also be the density preferred by the jurisdiction. The possibility of giving more development potential to one acre of receiving area versus another was not a problem in Ave Maria because a single partnership is developing the entire town.

Ave Maria is a mixed-use community planned for 11,000 residential units, 1.7 million square feet of retail, office, and business park structures plus a variety of recreational facilities. As of 2020, the estimated population was 10,000, including enrollment at Ave Maria Catholic University, which alone is expected to ultimately accommodate 5,000 students. Developers are assured of their ability to use RLS credits and achieve the densities approved in the Ave Maria plan. Furthermore, there are no mechanisms allowing circumvention of the TDR requirements.

To date, Collier County's RLS has placed 54,962 acres of sending area land under easement, which is more gross acreage preservation than the amount achieved by many US TDR programs. The mere existence of Ave Maria was facilitated by the fact that the receiving area is surrounded by its own undeveloped sending area and consequently devoid of nearby residents that might be inclined to oppose a relatively compact new town development next door. However, critics observe that the remote location of Ave Maria conflicts with basic principles of growth management (Linkous & Chapin 2014). Another study argues that the method of sending site selection in this program does not result in an optimal outcome for sensitive environmental resources including panther habitat (Schwartz 2013).

Rural Fringe TDR Program - Collier County adopted a third TDR program in 2004 for its Rural Fringe Mixed Use (RFMU) district, a 73,222-acre area owned by approximately 10,000 different property owners. The Rural Fringe TDR program aims to protect sending lands containing large areas of wetlands, endangered species habitat and other significant environmental resources. The sending and receiving areas are mapped. The sending area contains 21,128 acres of land under private ownership and 20,407 acres held by public entities. The receiving area contains 22,020 acres. The RFMU district also has a neutral area with 9,667 acres.

In the sending areas, on site development is allowed at a density of one unit per 40 acres or one unit for every lot smaller than 40 acres created before 1999. Alternatively, sending area property owners can receive one Base TDR Credit per five acres placed under a qualifying easement. The county has set a minimum price of $25,000 per Base TDR Credit but bonus credits can be sold at any price negotiated between buyers and sellers.

For each Base Credit, sending site owners can gain one early entry bonus credit, an incentive designed to jump start the program, which was still in existence as of 2017, 13 years after the program launched. Sending area owners can also gain one Environmental Restoration and Maintenance Bonus TDR Credit when they elect to submit and implement an approved Restoration and Maintenance Plan. These plans must be prepared by a county-approved environmental contractor and are required to include a listed species management plan, a procedure to remove invasive vegetation, and financial assurance that the plan will be followed until the property has achieved sustainable ecological functionality or the property is conveyed to a public agency.

For each Base Credit, sending area property owners can obtain a third bonus credit known as a Conveyance Bonus TDR by conveying title in fee at no cost for a property with an approved Restoration and Maintenance Plan to an appropriate governmental agency. By allowing sending sites to gain all four levels of TDR credits, Collier County has created meaningful incentives for sending site property owners to choose preservation/restoration rather than on site development. As an example, the owner of a 40-acre sending site has the choice of building one dwelling unit on site or selling as many as 32 TDR credits.

Despite a large inventory of entitled properties and the difficulties of finding a conservation agency willing and able to own and manage sending sites, Collier County's RFMU TDR program had 7,347 acres of sending area land under recorded development right limitations as of June 2019.

Other wetland preservation TDR programs that have achieved notable success are profiled in Part III including: Central Pine Barrens, New York; Charlotte County, Florida; Lee County, Florida; New Jersey Pinelands, New Jersey; Palm Beach County, Florida; Sarasota County, Florida; and Tahoe

Regional Planning Agency, California/Nevada.

References

Hawken, Paul. 2017. Drawdown: The Most Comprehensive Plan Ever Proposed to Reverse Climate Change. New York: Penguin Books.

Linkous, E. & T. Chapin. 2014. TDR Performance in Florida. Journal of the American Planning Association. Volume 80. Issue 3. Pages 253-267.

Miami-Dade County. 2017. Report Evaluating Existing and Potential Development Density Transfer Programs -Directive No. 152550. Memo to Board of County Commissioners dated January 23, 2017.

National Science Foundation. 2016. Everglades mangroves' carbon storage capacity worth billions. Accessed 7-13-21 at https://www.nsf.gov/discoveries/disc_summ.jsp?cntn_id=190254.

Schwartz, K. 2013. Panther Politics: Neo-liberalizing Nature in Southwest Florida. Environment and Planning. Volume 45, pages 2323-2343.

CHAPTER 5

Historic Landmarks – San Francisco, California and New York City, New York

The process of demolishing existing structures and replacing them with new buildings consumes vast amounts of energy and produces substantial GHG emissions by transporting debris, extracting raw materials, fabricating building components, delivering supplies, and assembling the new building. Conversely, preservation maintains the energy already embedded in existing buildings and avoids the additional GHG emissions resulting from new construction. According to the National Trust for Historic Preservation, from 10 to 80 years of operational savings in a new structure are needed to offset the climate-changing impacts resulting from the demolition and replacement process (National Trust for Historic Preservation 2011). A total of 66 of the TDR programs profiled in Part III aim to preserve historic resources, often in addition to other goals. San Francisco, California, and New York City, New York, discussed below, have two of the more active historic preservation TDR programs in the country.

San Francisco, California

San Francisco arguably has the most successful historic preservation TDR program in the nation. This accomplishment largely stems from observation of the factors common to programs that achieve good outcomes: receiving area baselines that developers want to exceed, sending area regulations that motivate property owners to participate, few ways to obtain bonus density other than TDR, and confidence in the ability to use TDR to gain additional floor area.

The TDR program here launched as part of San Francisco's 1985 Downtown Plan which reduced as-of-right FAR but allowed qualifying developments to exceed that lower baseline by transferring unused floor area from historic landmarks. Specifically, 253 structures were designated as architecturally significant and another 183 were deemed contributory. Unlike procedures used in some cities, these historic designations did not require the consent of the landmark owners.

In San Francisco, it is practically impossible to demolish a structure designated as a significant building (Category I and II), a limitation that motivates the owners of these properties to participate in the TDR program. There are fewer restrictions against demolishing contributory buildings (Category III or IV), but after the owners use the TDR process, the city prohibits demolition or alteration of these structures as with significant buildings. Buildings designated as individual landmarks can also become sending sites. Category V buildings that meet eligibility requirements can also participate in the TDR program. When these preservation lots are located in the C-3 zoning district, they are potential "transfer lots", the city's term for sending sites.

Parcels zoned P (public) can be transfer lots if they are owned by the City and County of San Francisco and located in a P district adjacent to the C-3 zone. Proceeds from the sale of TDRs from qualified sending sites in the P zone must be used to rehabilitate these public buildings.

The difference between the actual lot area on a transfer lot and the floor area allowed by the C-3 zoning is the TDR potential. When transfer lot owners choose to participate, TDR can be transferred on a one-to-one basis to "development lots", the city's term for receiving sites, at the ratio of one square of additional floor area per TDR.

The C-3 zone, which serves as the receiving area, has various sub-districts with different baseline densities. Baseline is 9 FAR in the C-3-O district and can double in intensity to FAR 18 using TDR. In contrast, baseline is FAR 6 in the C-3-R and C-3-G districts but can reach FAR 9 by using TDR, a 50 percent increase in intensity.

TDR cannot be transferred to any lot with a significant or contributory building. However, the Historic Preservation Commission can grant an exception if the additional space transferred to such a receiving site is essential to making it economically feasible to retrofit a significant or contributory building to meet seismic safety standards.

TDR only affects the allowed FAR on a development site and not other development standards including building height, bulk, setback, sunlight access, and separation between towers. However, the height limitations have not been reported to be a hindrance to the use of TDR in San Francisco.

TDR are conveyed by certificates of transfer prepared and recorded by the city's zoning administrator. These certificates incorporate notices stating that the development potential of the transfer lot has been permanently reduced by the square feet of floor area transferred. Sending site owners can transfer all of their TDRs at one time or transfer portions of them incrementally. However, after the first TDR transfer, the owner must prepare and implement a restoration plan, which may include necessary seismic retrofitting for the landmark. The owners must spend some of the proceeds of TDR sales on restoration, but there is no minimum spending requirement, and the city allows considerable flexibility in these plans (Frye 2017). Once TDRs have been transferred, a significant or contributory building cannot be demolished unless the city determines that the building has lost all reasonable use or is an imminent safety threat.

Transfer lot owners can transfer TDRs directly to the developers of receiving sites or to intermediaries. As of 2017, neither transfers nor the TDRs themselves were taxed, a feature that generated a great deal of intermediary participation in the past (Frye 2017). Because the TDR program has been stable and reliable for decades, the development community and real estate professionals have been able to find TDRs and negotiate a mutually agreeable price on their own. The city decided not to create a bank in order to avoid competing with private interests and otherwise interjecting a public component into a functioning private market.

TDR is the only way to exceed baseline FAR in the C-3 with the exception of affordable housing. This largely explains why this program has succeeded while TDR often languishes in places where developers can obtain bonus development potential by building on-site amenities like locker rooms and eco roofs that yield additional floor area at a cheaper price than floor area transferred from offsite sending properties. However, the city has exempted at least one major project from TDR requirements: in 2006, the city approved a development agreement that waived the TDR requirement for the Trinity Plaza project, which otherwise would have been required to acquire 879,000 square feet of transferable floor area (Seifel 2013).

The application of TDR to a receiving site is allowed without discretionary approval. Even though other aspects of a development may still require public hearings, the ministerial nature of transferring FAR is also one of the reasons for the success of this program.

In 2013, San Francisco commissioned a study of its TDR program. At that time, 5.3 million square feet of TDR had been transferred, representing the preservation of 112 sending site landmarks. Another 2.3 million square feet had been certified at that time but not used and an additional 2.7 million square feet existed on sending sites that had not yet been certified. Despite an apparent abundance of supply, the consultant believed that much of these untapped TDRs were not readily available because they are in small amounts or constrained by ownership and transactional issues. For this reason, the report recommended certifying 1.2 million of the total 3.6 million square feet held in publicly owned buildings such as the Opera House and Veterans Building. The report further recommended selling these public TDRs for a minimum price of $25 per TDR (subject to annual review) the market price of each square foot of transferred floor area in 2013 (Seifel 2013).

Section 128 Transfer of Development Rights in C-3 Districts has been amended at least seven times, most recently in 2013 and 2015. Early in the program, TDRs could only transfer between sending and receiving sites in the same C-3 zone. As of 2013, the code allows the following four transfer options: 1) between sending and receiving sites throughout the C-3; 2) from transfer lots containing significant buildings in the South of Market Extended Preservation District to development lots in the C-3; 3) from transfer lots zoned P and adjacent to the C-3 to development lots in the C-3; and 4) from transfer lots with individually-designated landmarks in any C-3 district to development lots located in any C-3 district.

In 2018, San Francisco adopted Sec. 128.1 Transfer of Development Rights in the Central SoMa Special Use District. This code section aims to facilitate the economic viability of buildings of civic importance in the South of Market (SoMa) Special Use District. Sites with significant or contributory buildings in Central SoMa may become transfer lots for the purpose of transferring TDRs to development lots within the Central SoMa Special Use District. The TDRs available for transfer are calculated as the difference between the actual floor area of the landmark and its allowable gross floor area. This code section establishes allowable gross floor area for buildings in five height districts ranging from FAR 3 in height districts of 40 to 49 feet to a high of FAR 7.5 for projects in height districts over 85 feet. In addition to historic landmarks, buildings consisting completely of affordable housing units can qualify to become transfer lots in the Central SoMa TDR program.

New York City, New York

New York maintains two citywide transfer programs plus transfer mechanisms in ten special districts. Between 2003 and 2011, these TDR programs generated at least 421 development rights transactions with a value of more than $1 billion (Furman 2013). The four TDR programs designed to preserve historic resources are discussed here while profiles of all New York City TDR programs can be found in Part III.

Landmarks TDR Program: 74-79 – The Landmarks TDR program is often referred to as "74-79" for its section in the city Zoning Resolution. This mechanism was adopted as part of the city's 1968 Landmarks Preservation Law. It allowed, and still allows unused density to be transferred from landmark buildings to adjacent sites in order to protect historic buildings from inappropriate alterations or demolition yet provide a form of compensation to the landmark owner for these restrictions.

The city Landmarks Commission designated Grand Central Station as a landmark and denied the owner, Penn Central Transportation Company, permission to build an office tower on top of this building. The ensuing Penn Central lawsuit was ultimately decided in 1978 by the U.S. Supreme Court in favor of the city. The court found that the city's denial was not a regulatory taking under the Fifth Amendment of the US Constitution prohibiting governments from taking private property for public use without 'just compensation.' In its 1922 *Pennsylvania Coal Co. v Mahon* decision, the court found that a regulation can be a taking if it goes too far. But in *Penn Central*, the court did not conclude that a regulatory taking had occurred. Plus, the decision added: "…while these [development] rights may well not have constituted 'just compensation' if a taking had occurred, the rights nevertheless undoubtedly mitigate whatever financial burdens the law has imposed on the appellants and, for that reason, are to be taken into account in considering the impact of regulation." The US Supreme Court has yet to decide the question of whether or not TDRs constitute 'just compensation' in the event of an actual taking. However, the court's recognition of TDR as a legitimate mitigation measure gave communities throughout the US the confidence to adopt TDR programs.

The 74-79 mechanism allows the difference between the floor area of a landmark and the floor area that would be allowed by zoning (if it were not a landmark) to be transferred to adjacent zoning lots. In 74-79, 'adjacency' means not just abutting lots but lots that are across streets and intersections and connected via chains of lots under common ownership. Despite this generous definition of adjacency, the Landmarks TDR program generated transfers from only 12 landmarks between 1968 and 2013 including Grand Central Terminal, Amster Yard, India House, John Street Methodist Church, Old Slip Police Station, 55 Wall Street, Rockefeller Center, Tiffany Building, Seagram Building, University Club, and St. Thomas Church for a total transfer of 1,994,137 square feet. An additional transfer between parts of Rockefeller Center was approved but not built (NYC 2015).

Observers cite the following reasons for reluctance to use 74-79.
1) The bonus on the receiving site is limited to 20 percent of the receiving site's as-of-right floor area.
2) The sending site is subject to a landmark maintenance plan.
3) The Special Permit needed for approval is time consuming and expensive (estimated by one expert to cost $750,000 (NYC, 2015).
4) Many landmarks have little or no unused floor area to transfer. Although New York has 1,300 landmarks, only 466 had unused development potential as of 2015 and many of these landmarks have so little unused potential floor area that transfers are functionally infeasible.

5) All three types of TDR mechanisms are hamstrung because TDR changes FAR limits while often not relaxing other development requirements, such as building height and setbacks, which also define the achievable building envelope. Developers can seek exceptions to these non-FAR requirements using the special permit or variance applications but these processes are strict.

6) New York State Law limits residential buildings to 12 FAR, a scale that many developments can reach by incorporating affordable housing under the city's Inclusionary Housing Program.

Consequently, developers prefer whenever possible to use the ZLM process, (detailed in the New York City profile in Part III), which is ministerial and therefore faster, cheaper and more certain. The numbers reflect this preference: between 2003 and 2011, 74-79 was used twice compared with 385 transfers occurring under the ZLM process (Furman 2013; Furman 2014; Gilmore 2013).

Theater District - In 1967, the city adopted a Special Theater District which used development bonuses that motivated the construction of five new theaters. However, the Theater District provisions were less successful at saving existing theaters. After losing several theaters, the city adopted a special permit process in 1982 that made it easier to transfer unused development potential from listed theaters and added a bonus for theater rehabilitation. However, the program was constrained by a limited receiving area. By 1988, the Theater Advisory Committee had grown the number of listed theaters to 30. To increase transfer options, the city allowed receiving sites in the Theater District outside the core to add one FAR of bulk in return for making a payment to preserve a theater within the core. This incentive proved inadequate. Between 1982 and 1998, only four transfers occurred yet theater owners had an inventory of over two million square feet of unused development potential (NYC 2015).

In 1998, the city adopted several reforms to its Theater District program.

- The receiving area was expanded, and listed theaters were allowed to transfer unused development potential anywhere in the Theater Subdistrict.
- The Special Permit requirement was replaced by a certification process allowing receiving area bonuses of either 20 percent or 44 percent FAR.
- Receiving sites were required to pay a surcharge with the proceeds used exclusively to fund theater preservation and use. Between 1998 and 2015, this surcharge increased from $10 to $17.60 per square foot of bonus floor area transferred to receiving sites.

To qualify for these bonuses, theater owners must covenant to maintain the theater for legitimate theater uses for the life of the receiving site development. TDR prices are determined by private transactions and in 2015 averaged about $225 per transferred square foot. Between 2001 and 2015, the Theater District became the city's most active TDR program, transferring roughly 500,000 square feet in 15 transfers from nine theaters (NYC 2015).

Grand Central Subdistrict – The 74-79 Landmarks Preservation TDR process transferred only 75,000 square feet from Grand Central Station between 1968 and 1992. In response, the city adopted the Grand Central Subdistrict of the Special Midtown District in 1992 to further the preservation of the Grand Central Station and other landmarks plus maintain pedestrianism and area character. In 1992, Grand Central Station was the only designated landmark with any appreciable amount of unused development potential, but it had a 1.7 million square feet of TDR supply. To motivate transfers, the

1992 changes allowed sending site TDRs to be transferred to receiving sites anywhere in the subdistrict to achieve one FAR of bonus floor area via a certification process. Also, by special permit, receiving sites in this subdistrict's core can achieve up to 21.6 FAR. Between 1992 and 2015, five transfers from Grand Central Station increased the total transfer from Grand Central Station to 488,036 square feet (NYC 2015).

South Street Seaport – The city created the South Street Seaport Subdistrict in 1972 to support the South Street Seaport Museum plus preserve and restore the historic Schermerhorn Row landmark buildings which were defaulting on mortgages and at risk of demolition. A TDR bank was created to immediately save the buildings and hold the TDRs for later use on designated South Street commercial area receiving site developments. Chase Manhattan Bank and Citibank accepted the TDRs as partial satisfaction of loan obligations. In contrast with most traditional historic preservation TDR programs, the allocation formula is the floor area of the sending site building or five times the lot area including the land area of adjacent abandoned streets. Transfers were conducted by the relatively easy process of certification. By 2013, all but 340,000 square feet out of a total supply of 1.4 million square feet had been transferred to six receiving site projects at prices ranging from $110 to $150 per square foot in 2007-2008 (NYC 2015).

Other Historic Preservation TDR Programs that have achieved notable success are profiled in Part III including: Los Angeles, California; Palo Alto, California; and Seattle, Washington. An additional 61 historic preservation TDR programs are also profiled in Part III: Arlington County, Virginia; Aspen, Colorado; Atlanta, Georgia; Austin, Texas; Birmingham, Michigan; Chattahoochee Hills, Georgia; Clarksdale, Arizona; Clearwater, Florida; Coral Gables, Florida; Dallas, Texas; Delray Beach, Florida; Denver, Colorado; Douglas County, Nevada; East Hampton, New York; Easthampton, Massachusetts; Exeter, Rhode Island; Gorham, Maine; Hatfield, Massachusetts; Hillsborough County, Florida; Huntington (Town), Suffolk County, New York; Iowa City, Iowa; Islip, New York; Ketchum, Idaho; Largo, Florida; Larimer County, Colorado; London Grove, Pennsylvania; Lysander (Town) Onondaga County, New York; Madison, Georgia; Miami, Florida; Middle Smithfield Township, Monroe County, Pennsylvania; Minneapolis, Minnesota; New Castle County, Delaware; Newport Beach, California; Oakland, California; Osceola County, Florida; Palmer, Massachusetts; Park City, Utah; Pasadena, California; Pima County, Arizona; Pitkin County, Colorado; Pittsburgh, Pennsylvania; Pocopson Township, Chester County, Pennsylvania; Portland, Oregon; Providence, Rhode Island; Redmond, Washington; San Diego, California; Santa Fe, New Mexico; South Middleton Township, Cumberland County, Pennsylvania; Southold Township, Suffolk County, New York; St. Petersburg, Florida; Tarpon Springs, Florida; Vancouver, Washington; Wareham, Massachusetts; Warrington (Town), Bucks County, Pennsylvania; Washington, DC; West Palm Beach, Florida; West Vincent Township, Chester County, Pennsylvania; and Westfield, Massachusetts.

References
Frye, Timothy. 2017. Presentation by San Francisco Historic Preservation Officer Timothy Frye, May 12, 2017, California Preservation Conference, Pasadena, California.

Furman Center for Real Estate & Urban Policy. 2013. Buying Sky: The Market for Transferable Development Rights in New York City. New York: New York University.

Furman Center for Real Estate & Urban Policy. 2014. Unlocking the Right to Build: Designing a More Flexible System for Transferring Development Rights. New York: New York University.

Gilmore, Kate. 2013. A Process Evaluation of New York City's Zoning Resolution (ZR) 74-79: Why Is It Being Used So Infrequently? New York: Columbia University.

National Trust for Historic Preservation. 2011. The Greenest Building: Quantifying the Environmental Value of Building Reuse. Accessed 7-14-21 at file:///C:/Users/Richard/Downloads/The_Greenest_Building_lowres.pdf.

New York City Department of City Planning (NYC). 2015. A Survey of Transferrable Developments Mechanisms in New York City. New York: Department of City Planning.

Seifel. 2013. TDR Study: San Francisco's Transfer of Development Rights Program. Prepared by Seifel Consulting Inc. for the San Francisco Planning Department.

Los Angeles used one of its TDR mechanisms to restore and expand its historic library.

CHAPTER 6

Adapt to Sea Level Rise – Ocean City, Maryland and Sarasota County, Florida

Almost 40 percent of Americans live in coastal areas that are or may be vulnerable to sea level rise (Lindsey 2021). Miami already experiences "sunny day flooding" on a routine basis, causing an estimated property value loss of $337 million from 2005 to 2017 (Cappucci 2019). The problem is so obvious that Miami voters passed a bond measure in 2017 authorizing $200 million for measures aimed at adapting to sea level rise (Smiley 2017). By the 2030s, scientists at NASA predict that increased coastal flooding will occur throughout the United States as sea level rise and climate change team up with added gravitational pull created by the lunar cycle (NASA 2021).

A dozen jurisdictions have TDR programs designed, at least in part, to protect coastal resources in general or, in some cases, to guard against coastal storms which will intensify and become more destructive as sea level rises. Two of these TDR programs are discussed below and the others are profiled in Part III.

Ocean City, Maryland

Ocean City is a coastal resort town of almost 7,000 people on a barrier island along Maryland's Atlantic shore. With proximity to Philadelphia, Wilmington, Baltimore and Washington, D.C., Ocean City welcomes eight million visitors annually, supporting a built environment of $8.6 billion and $1.4 billion in annual tourism revenue (Ocean City 2017; Schechtman and Brady 2013).

According to Ocean City's 2017 Hazard Mitigation Plan, the town is in a high vulnerability category for the impacts of sea level rise, including increased flooding and storm damage plus erosion of beaches. Numerous storms have caused damage throughout the town's history.

The Ocean City Beach Replenishment Project, launched in 1988 by the State of Maryland, Worcester County, Ocean City, and the U.S. Army Corps of Engineers (USACE), is the town's most significant hazard mitigation measure to date. These four governmental entities have a 50-year agreement committing the USACE to re-nourish the beach as needed, with USACE designing and managing the process as well as paying for 53 percent of the cost. This project was estimated to have prevented $600 million in storm damage as of 2017. The town's 2017 Hazard Mitigation Plan cites sea level rise as a primary reason for the town to continue this project and the plan identifies it as a high priority mitigation strategy. As detailed below, TDR was used to help implement Ocean City's beach nourishment/dune system project (Ocean City 2017; Schechtman and Brady 2013).

In 1972, Ocean City adopted a build-to, or Construction Control Line (CLL) that prevented development on beachfront lots between that line and the ocean. The town did not have the money to acquire the property or easements east of that line in the 1970s. However, the federal government required a property interest in this land as a precondition for the beach nourishment/dune system program to proceed (Schechtman and Brady 2013).

In 1993, the town adopted a TDR ordinance to secure these easements at minimal public cost. Owners of property east of the CLL, the Beach Transfer-Sending (BT-S) overlay district, had to register and transfer ownership to Ocean City or the State of Maryland by July 1994 in order to gain the benefits of beach nourishment and dune construction. In return, these owners were issued one TDR per 500 square feet. However, no time limit was placed on the transfer of these TDRs

The sending site owners are able to sell these TDRs to developers wanting a 25 percent density bonus in the town's Beach Transfer Receiving (BT-R) district, an overlay zone covering inland areas designated for higher density development in the town's comprehensive plan. Here, receiving site developers can exceed baseline density by up to 25 percent by buying one TDR for each bonus hotel room or two TDRs for each extra multiple-family residential dwelling unit.

Ocean City is a highly desirable real estate market and the town's zoning code produced a baseline that developers want to exceed. As of 2013, over 400 TDRs were transferred and only 70 development rights remained. The TDR program produced the beachfront control required by the federal government and saved the town millions of dollars in land acquisition expenses. The town kept the TDR program extremely simple and, other than administration, it cost the city almost nothing (Schechtman and Brady 2013).

Sarasota County, Florida

Sarasota County lies 50 miles south of Tampa on the Gulf Coast of Florida. Unlike Ocean City, Maryland, the TDR program in Sarasota County aims at implementing a range of goals including the protection of coastal resources and the reduction of human vulnerability to coastal storm surge.

The county adopted its first TDR program in 1981 aimed at preserving environmentally sensitive areas, agriculture, parcels of historic or archeological significance, and barrier islands as well as reducing the large number of substandard lots in antiquated subdivisions. This program preserved 8,200 acres with 8,169 of these acres purchased by the county itself as part of a legal settlement (Linkous & Chapin 2014).

In 1999, Sarasota County adopted a Conservation TDR program establishing a Conservation Sending Zone (CSZ) overlay covering category 1 and 2 storm surge zones, watercourses, and areas of special flood hazard, as well as ecologically significant areas. Annexations and regulatory changes eventually undermined many of the primary receiving areas, however the Conservation TDR program did succeed in protecting 730 acres by 2012 (Linkous 2012).

In 2002, Sarasota County adopted its 2050 comprehensive plan which added other permutations to the TDR program. As of February 2021, Sec. 124-103 of the Sarasota County Unified Development Code designated its Residential Sending Zone (RSZ) as land meeting at least one of the following criteria: 1) A platted subdivision which is non-conforming due to lot size, lack of paved streets or drainage, or other deficiencies; 2) Environmentally-sensitive areas including lands of high ecological value; 3) Areas that should be retained in agriculture, open space or other conservation use; 4) Parcels of historical or archeological significance; and 5) Parcels on a Barrier Island.

Land designated as a Conservation Sending Zone (CSZ), the program's other sending area, must be at least 500 acres when combined with its receiving site and meet one or more criteria: 1) Sites of High Ecological Value; 2) FEMA-designated areas of Special Flood Hazard; 3) Category 1 and 2 storm surge areas; or 4) a watercourse or slough system along with associated contiguous wetlands and mesic hammock areas potentially including up to a 200-foot buffer.

The density increases allowed via TDR vary depending on whether the receiving site is zoned Residential Receiving Zone (RRZ), High Density Residential Receiving Zone (HDRRZ), Future Urban Development (FUD) or Future Urban Residential Receiving Zone (FURRZ). For example, in the RRZ, TDRs can be used to increase dwelling units to 125 percent of the maximum density allowed by the underlying zoning or beyond 125-percent by special exception but not greater than 13 units per acre. As another example, in the HDRRZ, which is intended to implement the Comprehensive Plan in Town and Village Centers, the maximum density achievable with TDR is established by an adopted plan but cannot exceed 25 units per acre.

By 2012, the third generation of Sarasota County's TDR program had protected another 1,242 acres by transfers to a single receiving site development, bringing the total for all three programs to 10,172 acres protected (Linkous 2012).

Other Coastal TDR Programs aim to protect coastal resources and consequently reduce risk from sea level rise and coastal storms including: Bay County, Florida; Calvert County, Maryland; Charlotte County, Florida; Largo, Florida; Lee County, Florida; Monroe County, Florida; Oxnard, California; Pacifica, California; Pismo Beach, California; and Sarasota County, Florida.

References

Cappucci, M. 2019. Sea level rise is combining with other factors to regularly flood Miami. Accessed 7-14-21 at https://www.washingtonpost.com/weather/2019/08/08/analysis-sea-level-rise-is-combining-with-other-factors-regularly-flood-miami/.

Lindsey, R. 2021. Sea Level Rise. Accessed 7-14-21 at https://www.climate.gov/news-features/understanding-climate/climate-change-global-sea-level.

Linkous, E. The Use of Transfer of Development Rights to Manage Growth: The Adoption and Performance of Florida County TDR Programs. PhD dissertation. University of Pennsylvania.

Linkous, E. & T. Chapin. 2014. TDR Program Performance in Florida. Journal of the American Planning Association, 80:3, 253-267.

NASA. 2021. Study projects a surge in coastal flooding, starting in 2030s. Accessed 7-14-21 at https://www.jpl.nasa.gov/news/study-projects-a-surge-in-coastal-flooding-starting-in-2030s.

Ocean City. 2017. 2017 Hazard Mitigation Plan. Accessed 2-13-21 at http://oceancitymd.gov/pdf/OCHazardMitigation.pdf.

Schechtman, J. and Brady, M. 2013. Cost-Efficient Climate Change Adaptation in the North Atlantic. Accessed 7-20-19 at https://www.regions.noaa.gov/north-atlantic/wp-content/uploads/2013/07/CEANA-Final-V11.pdf.

Smiley, D. 2017. Miami gets $200 million to spend on sea rise as voters pass Miami Forever bond. Accessed 7-14-21 at https://www.miamiherald.com/news/politics-government/election/article183336291.html

SMART CLIMATE ACTION THROUGH TRANSFER OF DEVELOPMENT RIGHTS

CHAPTER 7

Adapt to Wildfire – Pitkin County, Colorado and Los Angeles County, California

Climate change is a major driver of the wildfires now taking a catastrophic toll on lives, land, and economic stability. Many US communities recognize the need to reduce human exposure to wildfire by limiting or eliminating the construction of homes in the wildland-urban interface (WUI). However, there are hurdles to actually accomplishing this goal including the expense of using public funds to pay for the preservation of private land in the WUI. Pitkin County, Colorado, Los Angeles County, California, and other US communities have used various tools including TDR to redirect growth away from rural land, thereby protecting environmental benefits as well as reducing human exposure to wildfire.

In 2004, almost 40 percent of US homes were located in the WUI where houses mix with wildland vegetation, making them increasingly vulnerable to wildfire (Radeloff et. al 2004). Wildfire threats in the WUI have been steadily worsening. An estimated 4.5 million homes lie in the WUI in California, a state that experienced five of its six biggest wildfires in 2020. In California as well as many other states, communities have responded to the increased danger by retrofitting homes, adopting better building and site design standards for new homes, and improving the ability of threatened populations to shelter in place or flee from approaching wildfires (UC Berkeley 2021).

A 2021 report prepared by the University of California Berkeley Center for Community Innovation studied three fire-impacted communities, including Paradise, California, where the Camp Fire killed 85 people, destroyed 19,000 buildings (95 percent of the town's total number of buildings) including 14,000 homes, displaced 40,000 residents, and burned 150,000 acres. Not surprisingly, the study found that reducing or eliminating development in the WUI would result in the greatest protection from wildfire in addition to other benefits such as the preservation of natural and working lands (with their potential to sequester carbon), access to outdoor recreation, more efficient provision and maintenance of public facilities, and reduced GHG emissions produced when people live in compact, energy-efficient communities as opposed to low density sprawl. This study also evaluated a scenario involving the creation of compact communities called resilience nodes within the most defensible parts of the WUI. While resilience nodes provide fewer benefits than the managed retreat scenario, they nevertheless perform better than the business-as-usual scenario in which low density development is built and rebuilt within the WUI. In addition to publicly funded buyouts and conservation easements, the report recommends consideration of TDR as one way of permanently preserving greenbelts in the WUI and the shifting of development to less vulnerable locations (UC Berkeley 2021).

Another 2021 study reports that Boulder, Colorado survived the 2012 Flagstaff Fire with no loss of lives or homes thanks to the 100,000-acre greenbelt that encircles the city, which was created by various preservation techniques including TDR (Greenbelt Alliance 2021). Since the TDR programs in Boulder County are discussed above, this chapter focuses on the other programs: Pitkin County, Colorado, which has been reducing wildfire risk and achieving many other benefits by strongly regulating and preserving its WUI with a suite of land use tools that includes an extremely innovative TDR program; and Los Angeles County, California, where a TDR program generated multiple benefits in the Malibu Coastal Zone including reduction of vulnerability to wildfire.

Pitkin County, Colorado

Pitkin County surrounds the affluent ski resort of Aspen in the mountainous center of Colorado. To keep growth from degrading environmental resources, the county aims to steer higher-density development into the Urban Growth Boundary (UGB) and allow mostly lower-density residential, agriculture/ranching, and open space outside the UGB. The county also anticipates that climate change will cause wildfires to continue increasing in size, intensity, and frequency over time. To reduce human exposure to fire and other disasters, the county's Hazard Mitigation Plan recognizes that the TDR program works with UGB policy by motivating the redirection of growth from hazard-prone areas and rural zones into places more suitable for development where it can benefit from infrastructure, emergency services, and public services.

In 1996, the Pitkin County Board created the county's Rural/Remote (R/R) zoning district to limit growth in backcountry areas which are characterized by sparse development, little or no access to traditional county emergency services, absence of utility districts, and exposure to natural hazards. This zone requires a minimum lot size of 35 acres and cabins here cannot exceed 1,000 square feet of floor area. Alternatively, owners of property in the R/R can choose to record covenants permanently precluding further development in return for one TDR per each full 35 acres or legally created parcel smaller than 35 acres.

Since 1996, Pitkin County has expanded opportunities for owners to participate by adding sending area potential for various forms of preservation in other rural districts including environmentally sensitive properties, properties designated in the Pitkin County Historic Register, and land with visual constraints or environmental hazards including geologic instability, steep slopes, and vulnerability to wildfire.

As detailed in the profile of Pitkin County in Part III, sending site allocation ratios are generally one TDR per 35 full acres in sending areas zoned R/R, CD-PUD, or Limited Development Conservation. The allocation is one TDR per ten full acres in the TR-1 and TR-2 zones and the Board of County Commissioners (BOCC) may also issue TDRs to constrained sites, visually constrained sites, and properties designated in the Pitkin County Historic Register.

The Pitkin County TDR program motivates owners of qualified sending sites to participate with sale prices that ranged from $115,000 to $318,000 per TDR between 2007 and 2019. In 2019, ten TDRs sold at prices ranging between $225,000 and $240,000 each (Condon 2020).

In receiving areas, property owners can use TDRs from any sending site to obtain an exemption for a new development right in the Aspen UGB under the Growth Management Quota System (GMQS). The GMQS uses a point system to score proposed developments based on how well they meet desired criteria. A proposed development that scores high may receive a permit to build a residential unit and/or gain bonus floor area within an individual dwelling unit through the GMQS annual quota without having to buy TDRs. However, a low-scoring project may have to wait year after year unless the developer uses the TDR option. The number of TDRs required for exemptions from GMQS varies depending on the size of the proposed residence. Transfers used to receive an exemption to build a new residential unit are approved by a One-Step Special Review by the Board of County Commissioners

TDRs from any sending site can also be used to exceed a baseline floor area of 5,750 square feet on receiving sites in the Aspen UGB and approved subdivisions up to the final maximum floor area allowed within each individual home. For receiving sites zoned TR-2, each TDR yields 1,000 square feet of additional floor area. On receiving sites in other zones, the ratio is 2,500 additional square feet of floor area per TDR. These transfers are approved by a One-Step Special Review conducted by the Hearing Officer.

As of 2020, the county had issued certificates for 389 TDRs and 254 TDRs had been transferred to receiving sites, with 82 percent of these TDRs used to exceed the baseline floor area of 5,750 square feet. At that time, TDRs had preserved a total of 8,879 acres.

Los Angeles County, California

Los Angeles County has had a TDR mechanism operating for decades within its Malibu Coastal Zone, which extends from the City of Los Angeles to the Ventura County border and roughly five miles inland from the shores of the Pacific Ocean. The Countywide Sustainability Plan adopted in 2019 recommends continued use of TDR as a way of steering development away from ecologically sensitive lands and areas most at risk from climate impacts including wildfire.

The Malibu Coastal Zone is riddled with antiquated subdivisions with small lots approved long before the adoption of modern standards for safety and environmental protection. In 1978, 64 percent of the 13,475 lots of record in the Malibu Coastal Zone were vacant. The Zone contains remarkable biodiversity including 900 plant species, over 50 percent of the bird species found in the country, and habitat for several charismatic species such as golden eagles, bobcats, and mountain lions.

The area has also been repeatedly damaged by drought-fueled wildfires, such as the 2018 Woolsey Fire which killed three people, cost $6 billion, destroyed 1,643 buildings, and scorched 88 percent of the Santa Monica Mountains National Recreation Area. The steep terrain has limited ability to absorb intense rainfall, particularly after being blackened during the wildfire season, leading to floods, debris flows, and landslides. Between 1992 and 1995, Malibu was declared a federal disaster area five times due to wildfires, floods and landslides.

The topography here resists the creation or expansion of urban infrastructure. This has made the Zone a difficult place for the effective deployment of fire/rescue vehicles. Existing narrow, twisting roadways can easily be blocked just as residents are trying to flee and emergency vehicles are attempting to access areas threatened by fires, floods, and mudflows.

Initially the California Coastal Commission adopted a policy of requiring one transferable development credit (TDC) for each new lot or dwelling unit allowed in the Malibu coastal plain, the receiving area. In 1981, the allocation of TDCs in the sending areas expanded to include resource areas and substandard lots. As detailed in the Los Angeles County profile in Part III, the allocation formulas are currently more complicated. But the overarching goals remain: to reduce development potential in antiquated subdivisions, environmentally sensitive land, and hazard-prone areas without increasing the total development capacity of the Malibu Coastal Zone.

Between 1987 and the incorporation of the City of Malibu in 1991, this TDR program was very active largely because of support from the California Coastal Conservancy. The Conservancy formed a TDC bank that bought TDCs in antiquated subdivisions and worked with Los Angeles County to permanently retire lots in property tax default. To further facilitate transfers, developers were allowed to make payments in lieu of actual TDCs which the Conservancy used to buy TDCs. The Coastal Commission also capitalized the work of the Mountains Restoration Trust, a non-profit satellite organization of the Conservancy that could purchase TDCs at below market rates using creative techniques that are not always available to a governmental agency. In addition, the Trust found itself accepting TDCs as donations from homeowners wanting charitable-donation tax benefits in exchange for scenic easements. Because these TDCs were acquired for little or no money, the cost of TDCs averaged roughly $20,000, which was roughly two percent of the price of new lots in the receiving area. In addition, the subdivision of 10-acre parcels into 2.5-acre lots in the highly desirable receiving area was capable of three- and four-fold increases of per acre value (Wiechec).

Other Wildfire Adaptation TDR Programs are profiled in Part III including: Calvert County, Maryland; King County, Washington; New Jersey Pinelands; and Summit County, Colorado.

References

Condon, S. 2020. Transferable development rights under scrutiny as Pitkin County sharpens growth management tools. Aspen Times: 10-10-20. Accessed 2-19-21 at Transferable development rights under scrutiny as Pitkin County sharpens growth management tools | AspenTimes.com.

Greenbelt Alliance. 2021. The Critical Role of Greenbelts in Wildfire Resilience. Accessed 7-12-21 at file:///C:/Users/Richard/Downloads/The-Critical-Role-of-Greenbelts-in-Wildfire-Resilience.pdf.

Radeloff, V., R. Hammer, S. Stwert, J Fried, S. Holcomb, and J. McKeefry. The Wildland Urban Interface in the United States. 2004. Ecological Applications. 15(3), pp.799-805.

UC Berkeley Center for Community Innovation. 2021, Rebuilding for a Resilient Recovery: Planning in California's Wildland Urban Interface. Project of Next 10. Accessed 7-12-21 at https://www.next10.org/publications/rebuilding-resilient.

Wiechec, Elizabeth. Transfer of Development in the Malibu Coastal Zone. Unpublished paper prepared for Mountains Restoration Trust by former Executive Director. http://www.amlegal.com/nxt/gateway.dll?f=templates&fn=default.htm&vid=amlegal:lapz_ca.

Los Angeles County uses TDR to protect the Santa Monica Mountains and reduce exposure to wildfires, floods, and mudflows.

CHAPTER 8

Biodiversity – Palm Beach County, FL and San Luis Obispo County, California

One million species are at risk of extinction due to climate change, land conversion, and other human-made impacts including deforestation, overfishing, poaching, and invasive species (Leahy 2019). Biodiversity is essential to food production, health, and ecosystem services. One third of our total food supply relies on a wide diversity of pollinators including bees, beetles, ladybugs, and dragonflies. The extinction of species reduces our chances of discovering new medications and vaccines, which often result from investigations of the natural world. Biodiversity is also key to ecosystem services such as the protection of water supply, water quality, storm water management, carbon sequestration, and the production of oxygen (Harper 2018).

A total of 169 of the TDR programs profiled in Part III are designed, in whole or in part, to protect natural resources, environmental quality, and open space in general. Many of the TDR programs discussed in Part I offer good examples of the preservation of natural areas and consequently biodiversity including Montgomery County, Maryland, King County, Washington, Collier County, Florida, the New Jersey Pinelands, and Miami-Dade County, Florida. To introduce additional programs, this chapter discusses the TDR programs in Palm Beach County, Florida, and San Luis Obispo County, California.

Palm Beach County, Florida

Palm Beach County is located on the Atlantic Coast roughly 60 miles north of Miami. After decades of losing environmentally significant land and agriculture, the county adopted a TDR program in 1993 that is successful largely because of its TDR bank. The TDR section of Palm Beach County's Land Development Code (5.G.3.) has been amended at least six times, most recently in 2019.

As of January 2021, Section 5.G.3.F lists the following eligible sending sites:
- Lands designated RR-20 (Rural Residential, one unit per 20 acres)
- Lands designated as priority acquisition sites by the Environmentally Sensitive Lands Acquisition Selection Committee (ESLASC) or the Conservation Land Acquisition Selection Committee (CLASC) that meet criteria for rarity, biological diversity or the presence of species listed as endangered, threatened, rare or of special concern
- Lands designated as Agricultural Reserve (AGR)
- Privately-owned lands designated Conservation (CON)
- Other environmental or agricultural sites determined by the County Board to be worthy of protection

Section 5G3F4 provides the following allocation ratios minus one TDR for each residential structure remaining on the sending site:
- RR-20: one TDR per five acres, (a four to one transfer ratio)
- AGR (Agricultural Reserve): one TDR per acre.
- Priority acquisition sites outside the Urban Suburban Tier: one TDR per five acres
- CON (Conservation): one dwelling unit per ten acres
- Eligible environmentally sensitive sites: maximum density permitted by the Future Land Use designation for the property

Potential sending areas are often subject to environmental constraints which can make it difficult for owners to build on eligible sending sites at the nominal densities permitted by the zoning code. Therefore, sending area property owners may be more motivated to choose the TDR option than these transfer ratios might suggest.

Other than code-conforming dwelling units that existed on the sending site prior to the TDR application, all other uses must comply with the restrictions imposed by the perpetual conservation easement needed to sever TDRs.

AGR sending sites cannot retain TDRs for future development. Other types of eligible sending sites can retain some TDRs, but the county reserves the right to determine the portion of the sending site that will be subject to the conservation easement in order to link environmentally sensitive land, agricultural land, and open space.

To facilitate transfers, the Palm Beach County Code created a TDR Bank to purchase, hold and transfer TDRs. Importantly, the code specifically states that TDRS are generated from environmentally sensitive lands purchased by the county through 1999. That provision allowed the bank to create, hold and sell TDRs severed from environmentally sensitive sites purchased by the county using proceeds from bond measures totaling $250 million approved by Palm Beach County voters in 1991 and 1999. As a result, Palm Beach County now has 34 nature preserves totaling over 31,000 acres of environmentally significant land qualified to generate TDRs. The TDRs from these properties were placed in the TDR Bank, creating an inventory of 9,000 TDRs (Pruetz 2012).

Developers have the option of buying TDRs from the owners of eligible non-public sending sites. However, sales from the county's TDR bank save developers the time and uncertainty of having to find willing private sellers and negotiate a price for TDRs. Between 2002 and 2009, the prices for

TDRs from the county TDR bank ranged from $10,000 to $50,000 each for bonus market-rate single family dwelling units. In fiscal year 2004-2005 alone, the TDR bank sold 435 TDRs. (For TDR bank prices in FY 2020-21, see table below). During housing booms, the sale of TDRs can be a meaningful revenue generator. In fiscal year 2004-05, the TDR bank charged $25,000 per TDR and sold 435 TDRs, generating over $10 million attributable entirely to the TDR program (Pruetz 2012). Revenue from the sale of bank TDRs is used for management and further acquisition of environmentally sensitive lands and wetlands.

Following amendments in 2011 or 2012, the Board of County Commissioners annually establishes the TDR bank sales price using median sales prices reported for the month of March by the Realtors Association of the Palm Beaches. Generally, the bank's TDR price for single family, market-rate bonus units is ten percent of the current median sales price for single family homes and the bank's TDR price for TDRs used in multifamily units is ten percent of the current median sales price for condo/multiple family units. However, various discounts are available, as depicted in the following table explaining the bank's TDR sales prices effective July 1, 2020 (Palm Beach County 2020).

Unit Type 1)	Full TDR Price (10% of Median Sales Price)	Workforce Housing Program (WHP) (5% of Full TDR Price)	Affordable Housing Program (AHP) (1% of Full TDR Price)
Single Family 2)	$37,000	$1,850	$370
Multi Family	$19,900	$995	$199
Unit Type 1)	Neighborhood Plan Price 3) (75% of Full TDR Price)	Neighborhood Plan WHP Price (5% of NHP TDR Price)	Neighborhood Plan AHP Price (1% of NHP TDR Price)
Single Family 2)	$27,750	$1,388	$278
Multi Family	$14,925	$746	$149
Unit Type 1)	Revitalization, Redevelopment, and Infill Overlay (RRIO) Price 4)	RRIO WHP Price (5% of RRIO TDR Price)	RRIO AHP Price (1% of RRIO TDR Price)
Single Family 2)	$9,250	$463	$93
Multi Family	$4,975	$249	$50

1. Per Article 5.G.3.G.4.c. TDRs shall proportionally reflect the unit mix of the non-TDR units
2. Single Family includes single-family detached, zero lot line, and townhouse
3. West Lake Worth Road Neighborhood Plan only, per Article 5.G.3.G.4.d.1) of the Unified Development Code
4. Per FLUE Policy 1.2.1-e and Article 5.G.3.G.4.d.2) of the Unified Land Development Code, includes Urban Redevelopment Area, Countywide Community Redevelopment Areas, and Lake Worth Park of Commerce Urban Redevelopment Area

Receiving projects can be located within sites where developers request increased density in Planned Development Districts (PDDs) and Traditional Development Districts (TDDs) or where developers propose subdivisions that exceed the density of their land use designations. Land eligible for sending sites cannot qualify as receiving sites unless the receiving project is providing affordable housing units.

Receiving sites must be within the Urban/Suburban Tier, consistent with county plans and codes, and compatible with surrounding land uses, particularly environmentally sensitive lands. To promote compatibility with environmentally sensitive areas, the code requires buffer zones with native vegetation ranging from 50 to 200 feet depending on the density of the receiving site development. Increased buffer zones are also required when receiving site projects with higher densities are located near certain lower-density zoning districts including zones for single-family residential lots of 14,000 square feet or greater.

Palm Beach County has various affordability requirements for bonus units generated by TDR. Specifically, 34 percent of all TDR bonus units must be Workforce Housing Program (WHP) units. When a receiving site project mixes WHP with Affordable Housing Program (AHP) units, the Planning Director determines how to apply the bonuses offered for WHP versus AHP units.

In an effort to shift density from the western to the eastern parts of the county, the Palm Beach County TDR program offers different maximum density bonuses via TDR. West of the Florida Turnpike, density bonus is limited to two units per acre versus three per acre east of the turnpike. Bonuses can reach four units per acre in Revitalization Redevelopment and Infill Overlay areas and additional variability occurs when a project uses WHP and AHP density bonuses. An additional unit of bonus density can be gained at receiving sites in the Urban/Suburban Tier within one quarter mile of a public park, mass transit facility, commercial facility, or major industrial facility. To avoid land use conflicts, density bonus cannot exceed 100 percent of baseline density for receiving site projects in low-density zoning districts (with some exceptions).

The Planning, Zoning and Building Department reported in its Transfer of Development Rights 2010 Annual Report that the Palm Beach County TDR bank sold 415 TDRs in FY 2007-08, 2008-09 and 2009-10 leaving an inventory of 6,919 TDRs in the bank at the end of FY 2009-10.

San Luis Obispo County, California

San Luis Obispo County lies 230 miles south of San Francisco and 190 miles north of Los Angeles on the Pacific Coast. The first of the county's two transferable development credit (TDC) programs was launched in part to preserve a rare tree species. This program is facilitated by a private non-profit conservancy, which by 2016 had purchased over 350 lots in the coastal community of Cambria using a small loan to launch an ongoing revolving fund for preservation (Johnson 2016). The Cambria TDR program is discussed in this chapter. The profile of the county's second program, which has preserved 5,464 acres, can be found in Part III.

The unincorporated community of Cambria, population 5,647 (2019), straddles scenic Highway One on the Pacific coast in northwest San Luis Obispo County. Cambria is home to a forest of rare Monterey pines. In the 1920s and 1930s, thousands of small lots were platted in the Lodge Hill subdivision, often on steep, highly erodible slopes. Many of these lots are still undeveloped and the county is gradually reducing their number using various means including TDC.

51

Before the certification of San Luis Obispo County's Local Coastal Program, the California Coastal Commission regulated development in the coastal zone. During this period, the Coastal Commission required developers to retire one small, steep lot in the Lodge Hill subdivision in return for approving a permit to build a home on another lot. This process became increasingly formalized from 1980 to 1988 with the development of the County's Local Coastal Program (LCP). The LCP ultimately targeted a canyon filled with Monterey pine, known as Special Project Area (SPA) #1 Fern Canyon and SPA #2, Visible Hillside (as viewed from Highway One), as two areas in which development should be minimized.

In 1988, the California Coastal Commission certified the Local Coastal Program for San Luis Obispo County and gave the county permit authority for the Cambria coastal zone. Some components of the Cambria TDC program appear in the county code at Section 23.04.440, including the basic mechanism of retiring environmentally sensitive lots in SPAs #1 and #2 and transferring the foregone allowable square footage to receiving sites in the Residential Single Family land use category within five small lot subdivisions in order to exceed baseline limits for building footprint and gross structural area. In one of this code's more unusual features, the program requires the participation of a non-profit corporation tasked with public information, program development, recordation of easements/deed restrictions, and the sale of available square footage from sending sites to the builders of receiving sites.

According to the county's North Coast Area Plan, TDC sending sites must meet at least one of the following seven criteria: 1) Located in Lodge Hill SPA #1 or #2; 2) Located in Monterey pine habitat and adjacent to an existing retired lot and containing at least one mature Monterey pine or four healthy saplings; 3); Containing steep slopes with known engineering problems or capable of causing adverse visual impacts from grading; 4) Areas where bluff erosion would necessitate shoreline protection; 5) Containing cultural resources; 6) Containing habitats for rare, endangered, or threatened species; and 7) Where development would cause adverse impacts when viewed from Highway One.

Eligible receiving sites are within the Cambria Urban Reserve Line, in any of five subdivisions, and served by sewer and water through the Cambria Community Services District. A receiving site cannot contain wetlands, habitat of rare or endangered species, identified cultural resources, slopes in excess of 30 percent, or where excessive grading or tree removal is proposed. Generally, lots in SPA #1 and #2 are intended as sending areas. However, lots in SPA#2 can receive TDCs transferred from sending sites that are also in SPA#2. Whether or not they receive TDCs, all development in the Highway One viewshed must be constructed with natural-looking materials in earthen or forest-toned colors.

Owners of qualified receiving sites may be required to purchase TDCs to offset development impacts. But primarily, the owners of receiving sites can choose to buy TDCs to gain increases in the allowable footprint and gross structural area of the property. The footprint is the lot area covered by residential and accessory structures including living areas, garages, and carports but excluding eaves, balconies and open decks. Gross structural area is the floor area within each structure including living space, garages, and carports but excluding exterior open decks and interior mezzanines.

Baseline for footprint and gross structural area varies depending on lot size and which of the following potential lot characteristics apply: 1) Special Project Area; 2) Forested Lot (containing at least one mature Monterey pine but not in an SPA or Marine Terrace; 3) Marine Terrace (an area between specified streets and the coastal bluff); 4) Steep Lot (30 percent or more); and 5) Typical Lot (not steep or in Marine Terrace and containing no Monterey pines).

In the lot size category of 1,750 to 3,499 square feet, a 1,750 square foot Forested Lot would have a baseline footprint of 600 square feet and structural area of 900 square feet and could use TDC to gain up to 100 square feet of additional footprint or structural area. In the lot size category of 3,500 to 5,249 square feet, a 3,500 square foot Forested Lot would have a baseline footprint of 1,200 square feet and structural area of 1,800 square feet and could use TDC to gain up to 300 square feet of additional footprint or structural area. In the category of lot sizes 5,250 square feet and larger, a 5,250 square foot Typical Lot would be allowed a footprint of 2,200 square feet for a one-story structure or 1,700 square feet for a two-story structure and a gross structural area of 2,200 square feet for a one-story structure or 2,600 square feet for a two-story structure; by using TDC, a receiving site with these characteristics could gain up to 400 square feet of additional footprint or structural area.

To comply with TDC requirements, receiving site owners buy transferable square footage from the Land Conservancy of San Luis Obispo County (Land Conservancy), the TDC program administrator. The purchase price must be at least sufficient to buy sending site easements and sewer assessments plus cover administration costs. The Land Conservancy bought the first sending area easements with a $275,000 loan from the California Coastal Conservancy. The Land Conservancy uses the proceeds from the sale of the square footage from these deed-restricted lots to purchase more sending site lots, thereby creating an ongoing source of funds from a relatively modest amount of seed money. The Land Conservancy aims for a sale price that receiving site purchasers find acceptable yet ensures that the Land Conservancy can continue to buy future easements without having to seek additional public funding. In the 1990s, the Land Conservancy was buying TDCs at an average cost of $10 per square foot and selling them for $20 per square foot.

As of 2016, the Land Conservancy had purchased and permanently preserved over 350 lots and formed the Fern Canyon Preserve which can be traversed on the Henry Kluck Memorial Trail (Johnson 2016). San Luis Obispo County doubled down on TDC in 2018 by approving an expansion of the program.

Other Biodiversity, Environmental, and Open Space TDR Programs that have achieved notable success are profiled in Part III including: Boulder County, Colorado; Calvert County, Maryland; Central Pine Barrens, New York; Charlotte County, Florida; Collier County, Florida; Douglas County, Nevada; King County, Washington; Larimer County, Colorado; Los Angeles County, California; Miami-Dade County, Florida; Montgomery County, Maryland; New Jersey Pinelands, New Jersey; Pitkin County, Colorado; Redmond, Washington; San Luis Obispo County, California; Sarasota County, Florida; Summit County, Colorado; Summit County, Utah; and Tahoe Regional Planning Agency, California/Nevada.

References
Harper, L. 2018. What is Biodiversity and how does Climate Change Affect It? Accessed 7-15-21 at https://news.climate.columbia.edu/2018/01/15/biodiversity-climate-change/.

Johnson, Jay. 2016. Request by the Department of Planning and Building for authorization to process updates to the Coastal Zone Land Use Ordinance and the North Coast Area Plan to expand the Cambria Transfer of Development Credits Program. Memo dated 11/15.2016 from Jay Johnson, Planning and Building to Board of Supervisors.

Leahy, S. 2019. One million species at risk of extinction, UN report warns. Accessed 7-15-21 at https://www.nationalgeographic.com/environment/article/ipbes-un-biodiversity-report-warns-one-million-species-at-risk.

Palm Beach County. 2020. Transfer of Development Rights (TDR) Program – TDR Bank 2020 Prices Effective July 1, 2020. Accessed 2-6-21 at PBC TDR Prices (pbcgov.org).

Pruetz, R. 2012. *Lasting Value: Open Space Planning and Preservation Successes.* Washington, D.C.: Island Press.

The City of Boulder and Boulder County, Colorado cooperated on the preservation of a greenbelt around the city using TDR and various other tools.

CHAPTER 9

Water – New Jersey Pinelands and Central Pine Barrens, New York

Roughly 1.6 billion people around the world live in countries already experiencing water scarcity, a number that the World Bank believes could double in the next 20 years (Kenney 2017). The United States is not immune from this problem as witnessed by the current drought that is threatening water supplies and exacerbating wildfires in western states. Decades ago, several US communities recognized the need to protect watersheds, critical aquifers, and groundwater recharge zones by reducing or eliminating development using a wide range of preservation tools including TDR. This chapter focuses on two TDR programs designed for multiple purposes including the protection of water supply and water quality: the New Jersey Pinelands and the Central Pine Barrens on Long Island, New York.

New Jersey Pinelands, New Jersey, roughly one million acres in size, encompasses the southeastern quarter of the State of New Jersey. The region is characterized by pine and oak forests, cedar and hardwood swamps, pitch pine lowlands, marshes, and bogs. The Pinelands is also home to the Pine Barrens tree frog, over 12,000 acres of "pygmy forest" (stands of dwarf pine and oak), 850 plant species, and more than 350 species of birds, reptiles, amphibians, and mammals.

The Pinelands region generates roughly one quarter of New Jersey's agricultural income, particularly cranberries and blueberries. It is also underlain by one of the largest and least polluted aquifers in the northeastern United States. In addition, it serves as an outdoor recreational area for the nearby New York City and Philadelphia metro areas, which are both just a one- to three-hour drive away.

The threat of encroaching development caused the U.S. Congress in 1978 to designate the New Jersey Pinelands as the first National Reserve and authorized the creation of a regional planning commission tasked with adopting a regional plan within 18 months. In 1979, the State of New Jersey adopted the Pinelands Protection Act which documented the goals of protecting the environment, safeguarding water quality, promoting compatible recreation/agricultural uses and nurturing appropriate development in the Pinelands. This act also endorsed the Pinelands Commission, which includes one representative from each of the seven counties in the region plus seven members appointed by the Governor of New Jersey and one member appointed by the U.S. Secretary of the Interior.

The Pinelands Comprehensive Plan, adopted in 1980, designated a 368,000-acre inner Preservation Area containing the most sensitive environmental resources. This is surrounded by a 566,000-acre Protection Area which had already experienced some development before 1980. The plan further identifies nine management areas.

The Pinelands Comprehensive Plan called for public acquisition of 100,000 acres using a combination of federal and state funds. However, the plan also relies on regulatory protection including a comprehensive TDR program requiring plan and code amendments from all of the local jurisdictions in the region.

Sending Areas - In the Pinelands TDR program, sending areas are located within the Preservation Area District, Agricultural Production Area and Special Agricultural Production Area. Jurisdictions were required to adopt land use regulations that promote preservation in these three areas. Codes were required to maximize open space protection using clustering provisions. Development could only be allowed by conditional use permit (CUP) here instead of as a matter of right.

Low density development can be approved by CUP in sending areas. But private owners are also motivated to preserve their land with a TDR transfer ratio allowing four development rights to be used at receiving sites in growth areas for every Pineland Development Credit (PDC) transferred from a sending area. In other words, the Plan offers four development rights at the receiving site for every Pinelands Development Credit (PDC) transferred from a sending site. The plan assigned 5,625 PDCs to the preservation areas which, at the four-to-one ratio, equates to 22,500 development rights in the Regional Growth Areas.

The plan establishes the number of PDCs available to a sending site based on development potential and environmental sensitivity. The Preservation District allocates one PDC per 39 acres of uplands, 0.2 PDCs per 39 acres of wetlands, two PDCs per 39 acres of land approved for mining but undisturbed, but no PDCs to land already mined. The Agricultural Production and Special Agricultural Production Areas allocate two PDCs per 39 acres for uplands, lands in active berry agriculture, wetlands in active field agriculture as of 1979 and uplands approved for mining but not yet disturbed, 0.2 PDCs per 39 acres for other wetlands, and no PDCs to uplands already mined.

The TDR program was initially facilitated by the Burlington County Conservation Easement and Pinelands Development Credit Exchange, essentially a TDR bank established by Burlington County, one of the seven counties within the Pinelands, using a $1.5-million county bond. The Exchange purchases PDCs as a buyer of last resort from sending sites in Burlington County but will sell its PDC holdings for use on receiving sites anywhere in the Pinelands.

In the early 1980s, the Exchange bought and sold PDCs at $10,000 each, which had the effect of establishing the price of PDCs in private transactions. In 1987, the State of New Jersey created the New Jersey Pinelands Development Credit Bank (NJPDCB) with $5 million from the state general fund. The NJPDCB acts as a "buyer of last resort" for PDC sellers unable to find a buyer. In an effort to keep bank purchases from interfering with the private PDC market, the NJPDCB cannot pay more than 80 percent of market value for PDCs. The NJPDCB must be re-authorized to buy PDCs every two years. The sale of PDCs from the NJPDCB cannot hamper private sales. Its PBCs are sold at auction. As required by the State of New Jersey, the minimum bid was initially set at $10,000 per PBC or $2,500 for development right. At the first auction, in 1990, the highest bid was more than twice that amount: $5,560 per development right (or $22,240 per PDC). The minimum bid can be reset as needed by State of New Jersey.

The NJPDCB can also award PDCs at no cost to projects that fill a compelling public need if the bank board finds that the conveyance is essential for the project to proceed and that the conveyance will not significantly impact the private PDC market.

The majority of PDC transactions occur on the private market. Between 1983 and 2020, 3,899 development rights were purchased by private parties while banks purchased 2,594 rights, or roughly 40 percent of the total of 6,493 rights.

In addition to buying and selling development rights and providing credit guarantees, the Pinelands Development Credit Bank:

- Guarantees loans secured by PDCs as collateral;
- Facilitates all PDC transactions;
- Issues PDC Certificates;
- Reissues Certificates when PDC ownership changes;
- Maintains the Registry of all PDC transactions;
- Uses the Registry to help PDC buyers find PDC sellers;
- Maintains a list of developers who want to buy PDCs; and
- Prepares an annual report of all PDC transactions.

Receiving Areas - Receiving areas are located in the Regional Growth Area, outside of the Core. The plan establishes Regional Growth Areas in 22 municipalities as suitable for and capable of accepting up to 46,200 transferred units. This is more than twice the number of units which would be generated by the severing of all credits allocated to the sending areas: 22,500.

The plans and codes for each jurisdiction with a Regional Growth Area designation establish the baselines and maximum densities for the receiving zones tasked with accommodating transferred PDCs. The PDCs can be used to boost single-family, multiple-family or any other kind of residential density. To assure developers of greater certainty in the use of PDC, the Regional Plan requires each jurisdiction to grant transferred density by right and not through a discretionary process.

In TDR programs, developers can sometimes circumvent compliance using rezonings, planned unit developments, and other processes that gain bonus density without the need to buy TDRs. To prevent this practice, all jurisdictions with receiving areas in the New Jersey Pinelands must require PDCs whenever they approve increased density or allow units in areas previously zoned for non-residential uses. In addition, the Pinelands Commission monitors local zoning codes and procedures for stringent development standards placed on applications for higher density which could have the effect of discouraging developers from buying development rights.

Program Evaluation and Modification - The Pinelands Commission has regularly studied its TDR program. In the 1980s, various issues hampered program success. 1) Transferred density was prohibited on land served only by septic systems. 2) Transfers were resisted by strong "no-growth" sentiments in some jurisdictions with receiving areas. 3) In some receiving areas, increased density was hindered by environmental constraints and/or height, setback and other development regulations that effectively discouraged transfers. 4) Duplicative regulations and disputes between local governments and the Pinelands Commission delayed transfers. 5) Landowners and developers perceived the transfer process to be complex and time consuming even though developers who employed the TDR option were typically interested in using it again.

In 1988, the Pinelands Commission instituted improved marketing, increased public education, streamlined approval processes, and establishment of the Pinelands Development Credit Bank. Adoption of the Pinelands Infrastructure Trust Bond Act helped finance roughly $50 million in sewer infrastructure in the Regional Growth Area, greatly facilitating the ability to increase density in the receiving areas. A decade later, New Jersey appropriated $3 million to buy and retire PDCs. In 2001, the state appropriated another $20 million to continue this Special Development Credit Purchase Program.

Carbon Sequestration - In 2006, New Jersey lawmakers mandated that the state reduce GHG emissions (using 2006 as a base year) 80 percent by the year 2050. In addition to emission reductions, New Jersey is planning to meet that target by using the carbon sequestration provided by forests and other natural sinks to offset gross emissions. According to the state's 80x50 Report, released in 2020, four of the five pathways to sequestration in New Jersey require land conservation: reforestation, proactive forest management, conservation of agricultural lands, and natural land preservation. (The fifth pathway is salt marsh and sea grass restoration/enhancement.) The 80x50 Report stresses the need to reduce the loss of upland forests, cropland, grassland, and wetlands in order for these pathways to be successful. The report includes the Pinelands Development Credit program in a list of twelve initiatives that are instrumental to open land preservation (New Jersey 2020). As shown in the Statewide Greenhouse Gas Inventory available in 2021, the New Jersey Department of Environmental Protection projects forests and other carbon sinks to offset over half of GHG emissions in the year 2050 under its green scenario (New Jersey 2021).

Outcome - In its 2020 Annual Report, the Pinelands Development Credit Bank reported that the program had preserved 55,392 acres, representing a private sector investment of $54.9 million. Of this total, 24,301 acres, or almost 44 percent were preserved in the Preservation Area District with the remainder in the Agricultural Production Area and the Special Agricultural Production Area.

Central Pine Barrens, New York

The Central Pine Barrens lies in Suffolk County, New York, which includes the entire eastern end of Long Island. The Pine Barrens is the largest single undeveloped area on Long Island and is home to pitch pine and pine-oak forests, coastal plain ponds, marshes and streams which provide open space and various outdoor recreational opportunities. The area protects the largest concentration of endangered, threatened, and special concern plant and animal species in the State of New York, including dwarf pines. In addition, the Pine Barrens constitute the deep recharge area for one of the largest sources of groundwater in New York State, an aquifer that provides drinking water for 1.6 million people.

Originally 250,000 acres in size, the Pine Barrens has been reduced by development to a 102,500-acre area shared by the towns of Brookhaven, Riverhead and Southampton. In 1989, environmental groups sued these three towns and Suffolk County over the potential impact of more than 200 proposed building projects on the Pine Barrens. In 1992, the New York State Court of Appeals decided that a protection plan was needed, and the New York State Legislature subsequently adopted the Long Island Pine Barrens Protection Act of 1993. To implement the act, New York formed the Central Pine Barrens Joint Planning and Policy Commission with representatives from each of the three townships, Suffolk County, and the State of New York.

In 1995, the Central Pine Barrens Comprehensive Land Use Plan was adopted by the three towns and Suffolk County. The plan designated a 52,500-acre Core Preservation Area and a 48,500-acre Compatible Growth Area. In the Core, most forms of development are prohibited (with some exceptions). However, the owners of land in the Core can permanently forego development potential and sell the resulting Pine Barrens Credits (PBCs) to increase development potential at receiving sites in the Compatible Growth Area and other locations outside the Core. Transfers are approved as a matter of right when the sending and receiving sites are both within the same town. Interjurisdictional transfers can also occur but require approval from both jurisdictions.

The names and addresses of buyers and sellers of PBCs are enrolled on the Pine Barrens Registry maintained by the Pine Barrens Clearinghouse. Sending site owners wishing to sell PBCs can choose to sell their PBCs using a real estate broker or by consulting the lists of potential buyers provided by the Clearinghouse. Alternatively, sending site owners may sell their PBCs directly to the Clearinghouse. The Clearinghouse Board of Advisors establishes a purchase price for PBCs. The Clearinghouse was originally capitalized in 1995 with $5 million no-interest loan from the State Natural Resources Damages Account. By buying and selling PBCs over the next ten years, the Clearinghouse was able to pay back the $5 million and retain an additional $3.2 million to continue its operations. At the start of 2019, the Clearinghouse had approximately $2.5 million in available funds.

To implement the Pine Barrens Plan, each of the three towns had to adopt receiving areas in the Compatible Development Area where developers could use PBCs as a matter of right without the need for special permits. The plan established a goal of creating receiving areas capable of accommodating 2.5 times the theoretical supply of PBCs in each town's sending areas.

From program inception to 2019, one out of every five credits crossed from a sending site in one of the three towns to a receiving site in another jurisdiction including the three towns with sending sites, (Brookhaven, Riverhead, Southampton), five other towns (Babylon, East Hampton, Huntington, Islip, Smithtown), and seven villages (South Country, Patchogue, East Hampton, Hauppauge, Quogue, Southampton, Westhampton Beach). However, most transfers have been between sending and receiving sites in the same town as described in one example from Brookhaven.

The Town of Brookhaven allows PBCs to increase residential density within single-family, multi-family, planned retirement community, nursing home, and assisted living districts, as well as increase floor area within seven business and two industrial zoning districts. The Planning Board, subject to Town Board limitations, may also require PBCs as part of a rezoning that results in additional density or intensity. Using an example of an upzoning for a residential subdivision, one PBC per unit has been required for 20 percent of the bonus units resulting from an upzoning. In addition, many subdivision approvals here have required the retirement of PBCs in order to meet wastewater flow regulations (Suffolk County 2014).

Transfer of credits representing increments of allowable wastewater flow is one of the distinguishing characteristics of TDR programs here. The Pine Barrens TDR program as well as most of the other TDR programs operating in Suffolk County use regulations and guidelines established by the Suffolk County Department of Health Services (DHS) to protect the aquifer that provides drinking water to much of Long Island. DHS essentially has a baseline density for development occurring on land served by on-site wastewater disposal. This is done to keep groundwater and drinking water within the required limits of 10mg per liter of nitrogen. However, DHS allows these baseline densities to be exceeded by TDR (Suffolk County DHS 2014). Each TDR is equal to 300 gallons of wastewater flow per day and can be transferred to receiving sites using the following conversion ratios.

1 TDR = 300 gpd wastewater flow
1 detached single family housing unit
2 attached units with maximum 600 gross floor area (GFA) each
1.3 attached housing units up to 1,200 GFA each
3 Planned Retirement Community (PRC) units up to 600 GFA each
2 Attached PRC units greater than 600 GFA each
10,000 Square Feet (SF) of dry retail space
7,500 SF of general industrial space
5,000 SF of non-medical office space
3,000 SF of medical office space
2,000 SF of wet (deli/takeout) space
10 seats of full-service restaurant (Suffolk County 2014).

As of January 1, 2020, the Pine Barrens TDR program had preserved 988 parcels representing over 2,000 acres with private sales totaling $53.5 million in value. During 2018, the average per-credit sales price was $78,000 with individual prices ranging from $44,000 to $100,000 per credit (Central Pine Barrens Commission 2018). Additional preservation has occurred under separate TDR programs operated by the towns of East Hampton, Huntington, Islip, Riverhead, Smithtown, Southampton, and Southold as well as three programs managed by Suffolk County itself.

Other Water-Protection TDR Programs that have achieved notable success are profiled in Part III including: Collier County, Florida; King County, Washington; Miami-Dade County, Florida; Montgomery County, Maryland; South Lake Tahoe, California; and Tahoe Regional Planning Agency, California/Nevada.

References

Central Pine Barrens Commission. 2018. 2018 Report. Westhampton Beach, NY: Central Pine Barrens Commission.

Kenney. 2017. Climate Change, water Security, and U.S. National Security. Accessed 7-16-21 at https://www.americanprogress.org/issues/security/reports/2017/03/22/428918/climate-change-water-security-u-s-national-security/.

New Jersey. 2020. *New Jersey's Global Warming Response Act 80x50 Report: Evaluating Our Progress and Identifying Pathways to Reduce Emissions 80% by 2050.* Accessed 2-12-21 at nj-gwra-80x50-report-2020.pdf.

New Jersey. 2021. Statewide Greenhouse Gas Inventory. Accessed 2-12-21 at NJDEP-Air Quality, Energy & Sustainability.

Suffolk County. 2014. *Suffolk County Transfer of Development Rights (TDR) Study*. Hauppauge, New York: Suffolk County.

Suffolk County DHS. 2014. *Suffolk County Department of Health Services General Guidance Memorandum #27: Guidelines for Transfer of Development Rights and Pine Barrens Credits for Sanitary Density Credit*. Hauppauge, New York: Suffolk County Department of Health Services.

Austin, Texas uses TDR to provide affordable housing, preserve historic landmarks, and protect areas essential to water quality.

PART II

Making TDR Work

*T*DR has been in existence for over 50 years. More than 300 jurisdictions use TDR. It has been extremely successful in several communities but produced disappointing results in many others. Some critics claim that the TDR concept itself is problematic. Conversely, in an article written with Noah Standridge that was published in the *Journal of the American Planning Association*, we argue that TDR programs languish because they do not observe ten factors that characterize the most successful US TDR programs (Pruetz & Standridge 2009).

The article, entitled What Makes Transfer of Development Rights Work?, identified the ten characteristics that appear most often in the 20 most successful TDR programs in the US. Two of these factors, demand for bonus development and optimal receiving areas, were found in all 20 programs that had protected the most land as of 2009. Three factors were found in at least 15 of the 20 programs: sending area development constraints, few or no alternatives to TDR, and market incentives. The other five factors appeared less frequently and were considered helpful but not essential to program success: certainty of ability to use TDR, strong public support, program simplicity, promotion/facilitation, and TDR banks. The following discussion explores each of these ten success factors in the order of their importance according to how often they appeared in the 20 most successful TDR programs in 2009.

Success Factor One: Demand for Bonus Development

TDR programs will not generate transfers if receiving area developers have little or no interest in whatever development advantage, they receive by participating. Many programs motivate developers to participate by offering additional receiving site residential density for choosing the TDR option. But, as shown below, there are actually many incentives that a program can offer.

Added Residential Density – In many TDR programs, the zoning designation of a residential receiving area has a baseline density that can be achieved as a matter of right, and a higher maximum residential density that can only be reached when developers choose to use TDR. Developers may decide not to use the TDR option when they are accustomed to building at or below the baseline density. Planners sometimes cite this situation as a reason not to consider creating a TDR program. However, there are strategies that can be used to address this phenomenon.

Engage Developers in Design of the TDR Program - In some jurisdictions, developers build at or below baseline density largely because it is something they know how to do and feel certain they can reliably profit from. Planners have an opportunity in these instances to work with developers to demonstrate that they can generate even more profit by exceeding baseline even though the higher above-baseline density involves the extra cost of compliance with TDR requirements. Logically, developers will be more inclined to exceed baseline if they have been involved in the development of a TDR program. Chesterfield Township, Burlington County, New Jersey offers a good example. Rather than promoting low-density construction common in this rural community, Chesterfield planned its receiving area as a traditional neighborhood design (TND) development with a density comparable to the nearby historic village of Crosswicks and features like local-serving retail, offices, recreational sites, public facilities and an elementary school plus walking paths connecting the various neighborhoods with each other as well as Crosswicks. This attention to detail required extraordinary public involvement and active participation from the development community. By doing this legwork in advance, Chesterfield was able to adopt detailed development requirements and architectural design standards that eliminated the need for public hearings as long as builders followed the rules and transferred the required number of TDRs. For this and other reasons, development firms were primed to use the TDR option as indicated by the fact that almost the entire receiving area was quickly developed at the with-TDR densities.

Require TDRs for Upzonings – Planners sometimes claim that developers are not interested in exceeding the maximum densities allowed by current zoning even when they annually process dozens of developer applications for upzonings. Livermore, California adopted a TDR ordinance that simply created a receiving area whenever a general plan amendment resulted in an increase in maximum allowable residential density. The TDR ordinance established the maximum density of the old zoning as baseline density and the maximum density allowed by the new general plan designation as the maximum density. The TDR ordinance also established the compliance needed to exceed baseline and achieve maximum density. This approach relieves TDR ordinances of the potentially contentious process of designating receiving areas. It also emphasizes that TDR is not inducing increased density but rather using upzonings to achieve multiple planning goals. In addition, this process allows jurisdictions to continuously create new receiving areas without the need to amend the TDR ordinance.

Downzone Receiving Areas – When developers decline to exceed current receiving area density limits, some jurisdictions have adopted TDR programs in which the baseline density is lower than the as-of-right density allowed by zoning in effect prior to TDR program adoption. This approach requires substantial political will and is best used when there is general agreement that the maximum density of the old zoning is so generous that it is failing to achieve multiple planning objectives. A TDR program was launched as part of San Francisco's 1985 Downtown Plan which reduced as-of-right FAR but allowed qualifying developments to exceed that lower baseline by transferring unused floor area from historic landmarks. As of 2013, the San Francisco TDR program had transferred 5.3 million square feet of unused floor area potential from historic structures, representing the preservation of 112 sending site landmarks.

As an example, from a rural preservation TDR program, Calvert County, Maryland, used two downzonings between 1978 and 2003 that reduced baseline density from one unit per five acres to one unit per 20 acres. These downzonings increased the motivation of the owners of productive resource lands to become sending sites rather than develop to the relatively low baseline density allowed on site. The downzonings likewise encouraged owners of property that qualified to become receiving sites to use the TDR option in order to exceed the lower baselines allowed after these downzonings. Studies of this program attribute much of the success of the Calvert County program to these downzonings.

Transfer Floor Area Allowed within Individual Residential Units – Traditionally, residential TDR programs involve transfers of density using baselines expressed as dwelling units per acre and bonus development potential allowed for entire individual dwelling units. But a growing number of programs express baselines for floor area within an individual dwelling unit which can only be exceeded by using TDR. The Cambria program in San Luis Obispo County, California imposes different baselines for a dwelling footprint and floor area depending on the size of a receiving site lot. In the category of lot sizes 5,250 square feet and larger, a 5,250 square foot Typical Lot would be allowed a footprint of 2,200 square feet for a one-story structure or 1,700 square feet for a two-story structure and a gross structural area of 2,200 square feet for a one-story structure or 2,600 square feet for a two-story structure; by using TDC, a receiving site with these characteristics could gain up to 400 square feet of additional footprint or structural area. Using this mechanism, this program preserved over 350 lots and established a preserve for a rare tree species.

A TDC program launched by Boulder County in 2008 allows new homes to exceed a baseline floor area of 6,000 square feet at the ratio of 500 square feet of additional floor area per TDC to a second threshold of 1,500 square feet above which two TDCs are needed per 500 square feet of additional floor area. In this program, the smaller the sending site house the more TDCs can be transferred up to a maximum of ten TDCs for legal building lots permanently restricted from any development.

The TDR program in Pitkin County, Colorado, which has preserved 8,879 acres to date, establishes a receiving area baseline of 5,750 square feet of floor area within an individual dwelling unit. This baseline can be exceeded at ratios ranging between 1,000 and 2,500 square feet per TDR depending on the zoning of the receiving site.

The TDR program in Aspen, Colorado, also uses floor area baselines within individual residential units. Owners of designated historic landmarks that have less floor area than baseline can sell one TDR per 250 square feet of unused floor area potential. Receiving site developers who choose the TDR option can exceed these baselines at the ratio of 250 square feet of floor area per TDR. Baselines vary between the seven zoning districts serving as receiving areas. On receiving sites zoned R-6, for example, single-family residential baselines range from 2,400 square feet of floor area on lots of 3,000 square feet or less up to 5,770 square feet of floor area within individual single-family residential units on 50,000-square foot lots. As of 2014, the program had created 64 TDRs and landed 24 TDRs at prices ranging from $175,000 to $240,000 per TDR.

Require TDR to Build Below as well as Above Baseline Residential Density TDR programs are sometimes seen as making it more difficult for developers to achieve the higher densities desired by jurisdictions. To resolve this conflict, at least two TDR programs require developers to retire TDRs when they build below as well as above a baseline density range. In Thurston County, Washington, the City of Olympia allows TDRs to be used to either increase density from seven units to eight units per acre or decrease density from five units to four units per acre in its R 4-8 Zoning District. According to the TDR ordinance in Orange County, Florida, in a tier with a baseline density of three units per acre, a 20-acre site could develop 60 units without TDR. However, 11 TDRs would be needed to increase density to 71 units or reduce baseline density to 49 units. This technique also appears in the TDR ordinance of Penn Township, Lancaster County, Pennsylvania.

Conversions of Residential Density into Other Forms of Development Potential– Receiving area developers may want additional development potential in the form of extra floor area, lot coverage, structure bulk, impervious surface coverage and building height or reductions in requirements for minimum development standards like lot size and setbacks. Here again, a robust public involvement process can help ensure that a TDR program ends up offering something that developers want. The profiles in Part III reveal dozens of conversions including the following.

Tahoe Regional Planning Agency recently revised its program, allowing conversions between single family residential units, multiple-family residential units, commercial floor area, and tourist accommodation units. In the TDR program in Queen Anne's County, Maryland, which had preserved 28,230 acres by 2016, TDRs can be used to increase residential density and net building area or reduce on-site open space requirements. Dozens of other examples can be found in Part III and at SmartPreservation.net.

In Hadley, Massachusetts, each TDR allows a bonus of an extra 2,000 square feet of commercial/industrial floor area plus the possibility of reducing lot coverage from a maximum of 30 percent to up to 70 percent and the ability to reduce the required parking area from 2.0 times to 1.5 times the building's floor area. Alternatively, developers may use one TDR to add two additional bedrooms to a senior housing development.

The TDR program in Warwick Township, Lancaster County, Pennsylvania, which has preserved over 1,600 acres of farmland to date, offers three types of incentives to developers of receiving sites. In the Campus Industrial Zone, developments can exceed a baseline of 10 percent lot coverage and achieve a maximum lot coverage of 70 percent by using one TDR for each 4,000 square feet of additional lot coverage. Buildings in this zone can also exceed a baseline height of 45 feet and attain a maximum height of 65 feet by using one TDR for each 4,000 square feet of additional lot coverage that would have been needed if the building were limited to a 45-foot height limit. In a third receiving site mechanism, elderly housing developments in the R-3 Residential zone can exceed a baseline of five units per acre and achieve a maximum density of 14 units per acre using one TDR per bonus unit per acre.

In the receiving areas of Redmond, Washington, one TDR can accomplish any of the following: 8,712 square feet of additional floor area; substitution for a requirement to provide 8,712 square feet of park land; 8,712 square feet of additional lot coverage or impervious surface coverage (with the increase not greater than ten percent of the site); 8,712 square feet of additional floor area in a story exceeding baseline building height (with restrictions in some zones). In addition, one TDR can allow receiving site developers to add up to five stalls above Redmond's maximum parking standards not to exceed a total of five stalls per 1,000 square feet of floor area.

The TDR program in Kitsap County, Washington, offers traditional receiving areas located within urban growth areas and designated in the comprehensive plan. In addition, the board of county commissioners may require TDRs for urban growth boundary expansions, site specific comprehensive plan amendments and rezonings. Development rights purchased for a site-specific amendment may also count toward any future upzoning request within the new designation. By resolution, the county requires 1 TDR per rezoned acre and anywhere from one TDR to three TDRs per acre for comprehensive plan amendments depending on the pre- and post-amendment comprehensive plan designations.

Transfers of Other Forms of Development Potential – Some TDR programs transfer forms of development potential that are indirectly related to density, floor area, and other structural regulations, including vehicular trips, wastewater flows, impervious surface coverage, and ability to proceed within building permit quotas.

Vehicular Trip Transfers – In some jurisdictions, the TDR program aims to allow flexibility in the distribution of density/intensity while maintaining an overall development cap in order to stay within the maximum capacity of an infrastructure system. Four TDR programs in California allow vehicular trips to be transferred between properties to keep traffic within roadway capacity. In the Silicon Valley, the City of Cupertino used a trip-transfer TDR mechanism to maintain acceptable levels of service in its DeAnza/Stevens Creek commercial corridor and yet allow for concentrations of floor area needed for the growing tech industry. Specifically, the baseline was 16 one-way trips per acre in the PM peak hour. This baseline could be converted to various types of land uses using PM peak hour trip generation factors contained in the city's Development Intensity Manual: 0.75 trips per residential unit, 1.0 trip per 1,000 square feet of office space, 2.0 trips per 1,000 square feet of general retail floor area, 3.5 trips per 1,000 square feet of restaurant and so forth. Developments that did not generate 16 trips per hour could transfer their unused trip quota to sites wanting to exceed that baseline by conditional use permit (CUP). At its peak, trip rights in this program were selling for $50,000 each. Some developers acquired trip rights before they needed them in anticipation of rising value over time. After 40 transfers, development had reached the capacity for the entire corridor. Apple's 750,000 square foot research and development office park was made possible by the acquisition of 322 trip rights from three separate sending sites. Consequently, this program allowed a major employer to stay in Cupertino without overwhelming the capacity of the city's transportation network. Cupertino provided a model for using transfers to allow necessary flexibility in the location of development within finite infrastructure systems, a model that has since been emulated in other California cities including Burbank, Irvine, and El Segundo.

Wastewater Flow Transfers – In the towns that provide receiving areas for the TDR program for the Central Pine Barrens of Long Island, New York, transfers are guided by regulations and guidelines established by the Suffolk County Department of Health Services (DHS) to protect the aquifer that provides drinking water to much of Long Island. DHS essentially has a baseline density for development occurring on land served by on-site wastewater disposal. This is done to keep groundwater and drinking water within the required limits of 10mg per liter of nitrogen. DHS allows these baseline densities to be exceeded at the ratio of one TDR per 300 gallons of wastewater flow per day and can be transferred to receiving sites using 11 conversion ratios ranging from one bonus detached single family residential housing unit per TDR to ten seats in a full service restaurant per TDR. As of January 1, 2020, the Pine Barrens TDR program had preserved 988 parcels representing over 2,000 acres with private sales totaling $53.5 million in value.

Impervious Surface Coverage – In addition to its transfer program for residential units, commercial floor area, and tourist accommodation units, the Tahoe Regional Planning Agency (TRPA) has another transfer program that allows the transfer of lot coverage rights between properties in order to allow flexibility in land development while still protecting the water quality of Lake Tahoe. In some areas, up to 30 percent coverage might be allowable without creating degrading amounts of sediment runoff and erosion. Conversely, in the most sensitive sites of the Stream Environment Zone (SEZ), lot coverage might be confined to as little as one percent of a site's total land area. These coverage regulations can constrain the ability to build new structures or expand existing buildings. As mitigation, TRPA developed a land coverage transfer program offering property owners the option of buying coverage rights from sending sites that permanently preclude excess coverage subject to limitations detailed in the TRPA profile in Part III.

Ability to Proceed within Building Permit Quotas – Several programs motivate developers to acquire TDRs in order to be able to build on schedule rather than waiting to get a building permit under various building allocation limitations. In order to keep growth from overwhelming the capacity of public infrastructure and services, TRPA sets annual limits on the amount of development allowed in the basin. For example, the quota might be 300 dwelling units, 400,000 square feet of commercial development, and 200 rooms of tourist lodging units. In one of several transfer mechanisms found in TRPA's program, a development can avoid the waiting list and build a new dwelling unit by using an allocation from a sending site that is vacant, has a land capability classification that is so sensitive the site is ineligible for development, and is permanently precluded from development by either deed restriction or ownership by a public or private non-profit agency tasked with open s pace preservation.

Similarly, Livermore, California motivated developer participation by reserving permit allocations within the city's permit quota system for receiving area projects that retired TDCs representing the preservation of land in the North Livermore sending area.

In receiving areas under the Pitkin County, Colorado TDR program, property owners can use TDRs from any sending site to obtain an exemption for a new development right in the Aspen UGB under the Growth Management Quota System (GMQS). The GMQS uses a point system to score proposed developments based on how well they meet desired criteria. A proposed development that scores high may receive a permit to build a residential unit and/or gain bonus floor area within an individual dwelling unit through the GMQS annual quota without having to buy TDRs. However, a low-scoring project may have to wait year after year unless the developer uses the TDR option. The number of TDRs required for exemptions from GMQS varies depending on the size of the proposed residence.

Groton, Massachusetts allows developers to use TDRs either to increase receiving site density or to build up to six extra dwelling units per year in addition to the six dwelling units per subdivision per year that would otherwise be permitted under the town's development permit limit provisions. Between 1988 and 2002, TDR preserved an estimated 500 acres of land including farmland, land important to Groton's water supply, and the Nashua River Greenway. During that time, developers used TDRs exclusively to exceed the permit quotas rather than increase density and TDRs were used in every subdivision with more than six undeveloped lots.

Santa Barbara, California encourages property owners to demolish existing oversized buildings, replace them with new, code-compliant buildings, and transfer the foregone floor area to receiving sites not for the purpose of building extra intensity but to be able to avoid the city's annual growth limits.

Islamorada, Florida also encourages the use of TDR by exempting some projects that incorporate TDR from its Building Permit Allocation System.

Success Factor Two – Optimal Receiving Areas

Receiving areas should be planned for the level of development allowed under the TDR program. This entails receiving areas that have few or no environmental constraints including an infrastructure system and or plan capable of accommodating the growth anticipated by the TDR program. It is also helpful to have receiving areas where existing residents accept new development or receiving areas that are buffered from existing residents who might oppose new development despite the fact that the new development is planned.

Existing Urban Centers – Where TDR programs designate sending areas within existing urban centers, the receiving areas are often within those existing urban centers as seen in many TDR programs designed to preserve historic landmarks (San Francisco, California), retain affordable housing (Seattle, Washington), expand infrastructure (Los Angeles, California), and implement many other community goals (New York, New York). In contrast, this section deals with the transfer of TDRs from rural sending areas into existing urban centers including nodes of higher-density development and downtowns. Existing centers make appropriate receiving sites since they maintain the compact urban form that is an essential foundation of sustainable development. Existing centers already have urban infrastructure, although expansion and/or repair of that infrastructure may be needed. The current residents of existing urban centers are also presumably accustomed to higher-density development and potentially less likely to oppose TDR receiving site developments. Furthermore, some cities designate underutilized central locations as TDR receiving areas in order to use TDR to achieve their redevelopment as well as preservation goals. Designating underutilized downtowns can be tricky since jurisdictions often offer financial incentives to motivate developers to build in these areas. Nevertheless, some TDR programs have managed to create successful receiving areas in revitalizing downtowns including King County, Washington, and South Lake Tahoe, California.

King County, Washington has partnered with five cities, (Seattle, Bellevue, Issaquah, Sammamish, and Normandy Park), to approve interjurisdictional TDR transfers from rural land under county jurisdiction. The first rural to downtown transfers occurred in 2000, under an interlocal agreement in which TDRs from agricultural sending areas in the county could be used to gain a 30 percent height bonus for residential buildings in Seattle's Denny Triangle Urban Village, a downtown district in need of revitalization. Per the agreement, King County pledged to spend up to $500,000 on Denny Triangle amenities including green streets, pedestrian/bicycle improvements, transit facilities/incentives, open space, storm water management, or public art/street furniture. In 2013, Seattle and King County signed another interlocal agreement that marked the first use of a Washington law allowing tax increment financing to fund infrastructure only in cities that adopt TDR receiving areas capable of accommodating that city's fair share of TDRs from sending areas throughout three counties in the Puget Sound Region. This is the only way that cities in the State of Washington can use tax increment financing. Per this agreement, Seattle creates a receiving area for 800 TDRs from sending sites under county jurisdiction and King County pledges to dedicate up to $15.7 million of additional property tax revenue to pay for open space and transportation improvements in Seattle's South Lake Union and Downtown districts.

At a smaller scale, the City of South Lake Tahoe, California, restored sensitive environmental sending areas and used the resulting TDRs to revitalize a downtown marred at one time by non-conforming motels and commercial structures build over a half century ago in response to revised gambling regulations and the Winter Olympics of 1960. During the height of its downtown revitalization, the South Lake Tahoe Redevelopment Agency acquired many of these older, nonconforming properties. Some of these acquired properties were recycled for new development using modern environmental standards. In some instances, an acquired site was unsuitable for development and was restored to its natural state in order to promote water protection benefits that in turn protect the water quality of Lake Tahoe. Importantly, the cost of acquiring the marketable rights needed to build new tourist units was offset by sales of the rights that the redevelopment agency acquired.

New Urban Centers – In addition to transferring TDRs from rural sending areas into existing urban centers, some TDR programs transfer rural TDRs into new centers. New urban centers can be compact communities that reduce GHG emissions by curbing sprawl and locating everyday uses close enough that residents can access schools, shopping, recreation, and employment without the need of a car. However, it is essential that new urban centers are properly located in order to minimize transportation and other environmental impact. The TDR programs in Montgomery County, Maryland and Collier County, Florida both locate receiving areas in new urban centers.

The TDR program in Montgomery County, Maryland, was adopted to implement the county's 1964 plan, *On Wedges and Corridors*, which aimed to concentrate development along a spine served by major transportation systems and other infrastructure flanked by a 93,000-acre Agricultural Reserve. As described below, some of Montgomery County's receiving areas consisted of single-family residential subdivisions outside of the Ag Reserve. But other receiving areas called for more diverse and compact development within the planned growth corridor, including the communities of Germantown and Clarksburg. Consequently, TDRs transferred to these new urban areas preserved rural sending areas while also implementing the goal of concentrating growth in places that limit GHG emissions and other environmental impacts caused by sprawl. As described in Parts I and III, Montgomery County has protected over 72,000 acres of the Ag Reserve to date, of which over 52,000 acres were preserved by TDR.

The Rural Lands Stewardship (RLS) TDR program in Collier County, Florida, has preserved 54,962 acres of rural land to date entirely through transfers to the new urban center of Ave Maria, the only receiving area to date for this program. Ave Maria is planned and partly developed for 11,000 residential units, 1.7 million square feet of retail/office/business park floor area, various recreational facilities, and a Catholic university designed for 5,000 students. In addition to other factors described in the Collier County profile in Part III, the success of this program is partly due to the fact that Ave Maria is surrounded by its own rural sending area which is largely owned by the same landowners who initiated and implemented the RLS program. This isolation reduces the potential for objections from adjacent residents, a factor that sometimes hobbles the ability to build higher-density, compact receiving areas. The builders of Ave Maria have aimed for a stand-alone town where people can satisfy most of their everyday needs by walking, biking or golf cart. However, critics observe that some residents inevitably commute in carbon-emitting cars to jobs in Naples, 36 miles west of Ave Maria.

Receiving Sites in Other Jurisdictions – The best receiving areas are often in a different jurisdiction than the sending areas. For example, the ideal receiving areas for a county TDR program might be within the incorporated cities of that county. However, the city officials may be reluctant to accept TDRs representing the preservation of land located in a different jurisdiction. Some cities may prefer to use TDR to safeguard city resources rather than risk exhausting TDR demand on the preservation of county land. In addition, some city officials may not want to quarrel with city residents who object to the idea of compact development, particularly when they imagine themselves as disadvantaged by transferred development potential for the benefit of county residents. In some cities, the elected officials themselves are wary of compact development and may even have run for office on no-growth or slow-growth platforms. It goes beyond the scope of this section to reiterate the many benefits of transitioning from business-as-usual growth to the compact development models needed for climate action. However, the following sub-sections discuss ways that some TDR programs have used to overcome barriers to inter-jurisdictional transfers.

Voluntary Partnerships – In some TDR programs, cities have been motivated to accept TDRs from county sending sites by forming partnerships that create mutually advantageous outcomes. As discussed above, the interlocal agreements between King County and incorporated cities often involve a payment from King County to make specified improvements within the city's receiving area. In the 2013 agreement, King County pledged to dedicate up to $15.7 million of additional property tax revenue to pay for open space and transportation improvements in Seattle's South Lake Union and Downtown districts in return for Seattle agreeing to accept up to 800 TDRs from King County sending areas. Similarly, in a 2009 agreement with Bellevue, King County agreed to fund $750,000 in stream improvements within the city.

In addition to monetary inducements, jurisdictions with receiving areas can be encouraged to accept TDRs from a different jurisdiction because preservation of the designated sending area benefits both communities. Boulder County, Colorado has signed intergovernmental agreements (IGAs) with eight municipalities and the unincorporated town of Niwot. When a city agrees to accept TDRs under these agreements, the sending areas are typically land under county jurisdiction but close to the city limits so that sending area preservation can accomplish any number of city land use goals involving the protection of important environmental resources and the creation of greenbelts. In the Boulder Valley agreement between the City of Boulder and Boulder County, these jurisdictions committed to TDR transfers and various other planning/preservation tools in a process that ultimately resulted in the greenbelt that now surrounds the City of Boulder and has been credited with protecting the city from at least one wildfire.

In another example of interjurisdictional cooperation, Larimer County, Colorado, and the City of Fort Collins partnered to protect a ¼-mile natural resource buffer around the Fossil Creek Reservoir. By intergovernmental agreement, the city and county created a program that produced the benefits of an inter-jurisdictional transfer program without actually having to transfer density between jurisdictions. The county and the city jointly planned the receiving areas using city development standards with the understanding that the receiving sites, although in the county, would be annexed to Fort Collins following approval of projects that utilized the TDR option. In fact, the program required receiving site projects to apply for annexation prior to recordation of plats and these properties were annexed as soon as possible following transfer of the TDRs. Even though the receiving site projects were approved by Larimer County, Fort Collins served as a referral agency and conducted its own review of receiving site projects and associated infrastructure improvements.

Intergovernmental Cooperation Aided by State Incentives – State governments can motivate counties and cities to cooperate in interjurisdictional TDR programs by offering incentives like those created by the State of Washington. In the Puget Sound Region of Washington, cities that meet regional TDR program standards can use the Landscape Conservation and Local Infrastructure Program (LCLIP) to fund infrastructure using tax increment financing, a tool not otherwise available in the State of Washington.

The Puget Sound Regional Council (PSRC) calculated the total number of TDRs in a three-county region and allocated these TDRs to larger cities in these counties based on growth projections and other criteria. To qualify for LCLIP, a participating city must agree to accept at least 20 percent of its TDR allocation, adopt an infrastructure plan for the receiving area capable of accommodating these TDRs, and establish at least one infrastructure project area. Cities are able to receive a greater proportion of tax increment revenues by agreeing to accept more than 20 percent of the allocation.

In 2013, King County and Seattle adopted the first interlocal agreement that satisfies the regional program requirements and allows Seattle to use the tax increment financing provisions of LCLIP. For up to 25 years, the county will share 17.4 percent of the new property tax revenues generated in these two receiving areas. This shared tax revenue will help pay for roughly $17 million in Green Street improvements in Seattle including transit, bicycle, and pedestrian facilities.

State-Mandated Interjurisdictional Cooperation – In a few instances, state governments require counties and municipalities to transfer TDRs inter-jurisdictionally.

The threat of encroaching development caused the U.S. Congress in 1978 to designate the New Jersey Pinelands as the first National Reserve. The New Jersey Pinelands Commission, an agency of the State of New Jersey, adopted a plan in 1980 to preserve farmland, forests, sensitive natural areas, and water resources using various preservation tools including a TDR program establishing an inter-jurisdictional transfer mechanism for the seven counties and 56 municipalities in this one-million-acre region. To implement the plan, 22 municipalities were required to adopt TDR receiving areas capable of accommodating TDRs transferred from sending areas anywhere within the region. As discussed in the profile for this program (in the Water chapter of Part I), the New Jersey Pinelands TDR program alone had preserved 55,392 acres as of 2020. As an indication of the need for state-mandated interjurisdictional cooperation, a second TDR program aimed at protecting land in the New Jersey Highlands does not require municipalities to create receiving areas for its TDRs and, at last check, no municipality has yet volunteered to accept credits from the Highlands.

The Tahoe Regional Planning Agency (TRPA) is tasked with protecting and restoring the clarity of Lake Tahoe and the general environmental health of the lake's 207,000-acre watershed. TRPA, a bi-state agency, maintains a multi-faceted TDR program operating in the City of South Lake Tahoe, two California counties and three Nevada counties. Originally, inter-jurisdictional transfers of residential unit potential, commercial floor area, and tourist accommodation units were subject to the approval of the sending and receiving site jurisdictions, a process that was recognized as complex and costly. Program amendments in 2018 eliminated this requirement, greatly facilitating interjurisdictional transfers.

Other Interjurisdictional TDR Programs – As described in the program profiles in Part III, interjurisdictional TDR programs include: Boulder County, Colorado; Central Pine Barrens, New York; King County, Washington; Larimer County, Colorado; Livermore, California; Morgan Hill, California; Miami/Dade County, Florida; New Jersey Highlands, New Jersey; New Jersey Pinelands, New Jersey; Puget Sound Region, Washington; Tahoe Regional Planning Agency, California/Nevada; and Warwick Village, Orange County, New York.

Greenfield Receiving Sites – To minimize GHG-generating sprawl, TDR receiving areas should be located within existing urban centers. However, many jurisdictions designate sending areas on land that is currently undeveloped but planned for growth. Ideally, these greenfield sites will be developed according to smart growth principles, meaning compact, mixed-use neighborhoods where residents can reach schools, shopping, recreation, and jobs with little or no need for a car. Receiving sites that are developed at suburban densities generate fewer climate action benefits than smart growth but they have nevertheless been responsible for producing some of the most successful TDR programs from the standpoint of the amount of sending area land preserved. This track record is partly due to the fact that subdivision development on greenfields is the dominant form of growth in many communities. In many places, builders are familiar with this form of development and can charge comparable prices for single family detached homes that are all within a particular range of lot sizes. In that suburban range of densities, builders can readily calculate how much they can profitably spend on a TDR.

As discussed above, some receiving areas in the highly successful TDR program in Montgomery County, Maryland, are in the higher-density category appropriate for new urban centers such as |the R-10 zone with a baseline of 43.5 units per acre and a maximum with-TDR density of 100 units per acre. However, the receiving sites that used the bulk of the TDRs between 1980 and 2007 were in lower density zones like R-200, with an effective baseline of two units per acre and a maximum |density of five units per acres when three units per acre are achieved using TDR (Walls & McConnell 2007).

Rural Receiving Areas – Establishing TDR receiving sites in rural areas has a potential to harm the resources that the TDR program aims to protect. Even when the least environmentally sensitive places are designated as receiving sites, development in rural areas allows additional people in remote locations where they are more likely to need private cars to reach schools, shopping, entertainment, and work. However, a jurisdiction may rationalize that it will experience some level of rural development with or without a TDR program and that the amount of preservation achieved more than compensates for impacts generated by rural receiving areas. As described in the following paragraphs TDR programs with rural receiving sites have achieved a notable amount of preservation.

Blue Earth County, Minnesota, is mostly zoned with a baseline density of one residential unit per quarter-quarter section (40 acres). However, property owners can transfer development rights from one quarter-quarter section to a contiguous quarter-quarter section if approved by permanently restricting development on the quarter-quarter sending site. Since these transfers occur at a one-to-one ratio, the overall development level does not change and, ideally, the transfers result in preservation of the most productive agricultural land and the most sensitive environmental areas. Consequently, even though the receiving sites are in the Agricultural or Conservation zoning districts, transfers are achieving key goals of the Blue Earth County Land Use Plan of maintaining agricultural areas and protecting natural resources. To date, at least 5,000 acres have been protected by this program.

In the Agricultural zone in Rice County, Minnesota, the baseline density allows one dwelling unit per quarter-quarter section. In this rural district, property owners can permanently preclude development of a sending site and transfer the resulting TDR to create cluster developments and planned unit developments that can also be located in the Agricultural zone. As detailed in the profile in Part III, all 14 Rice County townships have used TDR and the program has preserved more than 5,862.

In 2004, Queen Anne's County, Maryland, added Critical Areas, an overlay zoning district with an on-site density limit of one unit per 20 acres, as sending areas in its TDR program. The receiving areas could also be located in the Critical Area overlay, with individual densities reaching one unit per five acres as long as the overall density of the Resource Conservation Area did not exceed one unit per 20 acres. Critical Area TDRs were in high demand because they could boost density in the desirable waterfront properties zoned Critical Area Overlay. The value of Critical Area TDR started at $35,000 each but rose as high as $265,000 each by 2005 as supply dwindled. By 2005, the county's TDR and Non-Contiguous Development tools had protected almost 10,000 acres. A 2007 study concluded that Queen Anne's County's programs demonstrate that rural receiving areas can create substantial demand for the additional density provided by TDR (McConnell, Walls & Kelly 2007).

Success Factor Three – Sending Area Development Constraints

Successful TDR programs typically motivate owners of sending area land to choose the TDR option with features that have the effect of encouraging owners to transfer development potential rather than using it on sending sites. As discussed in the following four subsections, sending area property owners can be motivated by physical constraints, density limitations, development regulations, and offsite requirements.

Physical Constraints – Sending area property owners can be motivated to choose the TDR option by physical constraints such as topography, soil suitability, distance to infrastructure, or simply site isolation. However, TDR programs can either strengthen or weaken these motivations depending on policy considerations.

Some communities may choose to reduce or even prohibit TDR generation on sending sites that are currently considered difficult or impossible to develop. This decision may be aimed at reducing the supply of TDRs and/or targeting the preservation of what are considered to be the highest priority resources or locations. These communities may offer fewer or no TDRs for the preservation of land with physical constraints such as steep slopes, floodplains, or wetlands. In fact, many programs require sending site property owners to submit sketch plans, sometimes with soil reports and other technical studies with a goal of estimating how much development opportunity actually exists on a specific sending site and then only issue TDRs for the number of dwelling units that the sending site can feasibly accommodate. These TDR programs are largely counteracting the ability of physical constraints to motivate permanent preservation.

Other jurisdictions take the opposite approach. These communities allow the transfer of TDRs from sending areas with little or no regard for physical constraints. This approach may partly be designed to keep the program as simple as possible. It may also reflect a desire to protect multiple resources rather than just a single resource. For example, Montgomery County, Maryland, purposely qualified all land in its Ag Reserve as eligible sending sites because it wanted to preserve the integrity of the entire landscape, meaning streams, wetlands, and riparian corridors as well as farmland partly because these environmental resources are valuable in their own right and partly because the sustainability of agriculture in the Ag Reserve depended on a primarily cohesive rural area rather than one that was interrupted by scattered developments.

Some TDR programs may also allow properties with physical constraints to qualify for TDR because human ingenuity has a habit of overcoming these constraints. Today, we find view homes on hillsides that were previously considered unbuildable. Properties that were precluded in the past because they could not support on-site waste disposal can ultimately be developed with the approval of new, on-site systems as well as sewerage. Lands precluded from development because they are designated as floodplains, riparian corridors, or wetlands, can conceivably become eligible when governments change and adopt new classification systems. Land that is not under near-term threat of development due to distance from infrastructure or urban centers may ultimately be on the urban fringe as development expands outward over time.

When TDR programs allow sending areas with physical constraints to be sending sites, it motivates their owners to choose the TDR option. It allows these owners to receive compensation for development potential that may not materialize for years if at all. In that regard, it can motivate owners just as much as the other factors discussed below including density limits, development regulations and off-site requirements. Of course, jurisdictions must decide for themselves whether this policy best implements their planning goals. Some communities may be satisfied if their TDR program saves land that does not appear to be under the greatest threat of near-term development, perhaps land located at the periphery of a town or county. In other jurisdictions, this result might not be acceptable because the goal is to preserve land under the greatest near-term threat.

Density Limitations - Many TDR programs underperform because the sending area density limitations (or minimum lot sizes) are not adequately implementing the community's goals for the sending area. For example, a community's general plan might designate the sending area for long-term agriculture but the zoning that purports to implement that goal might allow residential development at suburban densities of one unit per five, 2.5 or even one acre. This jurisdiction is at best sending mixed messages to the owners of land in this sending area. The general plan suggests that the jurisdiction wants to protect its local agricultural industry. But the density limit, regardless of its name, facilitates the conversion of farmland to subdivisions. Faced with this mixed message, landowners logically will have doubts about the long-term viability of commercial agriculture since a subdivision might appear next door inhabited by urbanites who have been known to complain about dust, noise, odor, pesticides, slow-moving farm vehicles and other activities common in agricultural areas. As discussed in the Montgomery County, Maryland profile in Part III, these sending area property owners could experience impermanence syndrome, a resignation that neighboring subdivisions are inevitable. This syndrome can lead farmers to stop making agricultural investments and improvements, which in turn results in lower productivity and upkeep. Ultimately, the fear of encroaching development becomes a self-fulfilling prophesy as the area loses its agricultural vitality.

Conversely, in successful TDR programs, sending area regulations typically implement the general plan goals for the sending area or at least do not cause sending area property owners to doubt the sincerity of the general plan designation. The most obvious indication of seriousness is the maximum density and/or minimum lot size allowed by the sending area zoning. However, other land use regulations and mitigation measures can also be effective in implementing general plan goals for the sending area as discussed below.

Montgomery County, Maryland offers a good example of the need to adopt zoning that actually implements planning goals. The county's 1964 plan called for green wedges, but the zoning allowed one- and two- acre lots in these wedges. As a result, the county lost thousands of acres to subdivisions and continued to see sprawl despite a subsequent downzoning to one-unit-per-five acres. In conjunction with the adoption of its TDR program in 1980, Montgomery County downzoned the 93,000-acre Agricultural Reserve to one unit per 25 acres, the minimum size considered necessary for a viable farm. In other words, the zoning finally matched the goals of Montgomery County's general plan. Montgomery County proceeded to permanently preserve over 72,000 acres of the Ag Reserve because the county observed this and most of the other nine TDR success factors discussed in Part II.

Similar to Montgomery County, most of the other successful TDR programs apply maximum densities that implement the goals for their sending areas. The Forest Production District, which is one of several zones in the sending area of King County, Washington, has a minimum lot area of 80 acres. In Rice County, Minnesota, sending sites are limited to one dwelling unit per quarter-quarter section, which generates participation particularly from interior quarter-quarter sections that are not located on roadways. Many of the sending areas in the New Jersey Pinelands are limited to one unit per 39 acres and development is only allowed by conditional use permit rather than as a matter of right. In the Rural/Remote Zone of Pitkin County, Colorado, minimum lot size is 35 acres.

In some communities, density limitations that implement sending area goals are in place before TDR program adoption. When that is not the case, public officials may have to contemplate a downzoning of the sending area as occurred in Montgomery County, Maryland. However, downzonings tend to be contentious. If political pressure is too great, these jurisdictions may prefer to forego a downzoning and motivate property owners to choose the TDR option by using a high ratio of TDR allocations as discussed in the section below entitled Success Factor 5, Market Incentives. By leaving permissive sending area density limitations in place and applying higher TDR allocations per acre, a community can reduce opposition that might jeopardize adoption of a TDR program. However, higher allocation ratios mean less preservation occurs per TDR because each TDR represents a smaller amount of protected land.

Development Regulations – Density limitations are the development regulation most prominently featured in TDR. However, other development regulations can strongly motivate sending area property owners to choose the TDR option. In the Rural/Remote Zone of Pitkin County, Colorado, cabins in the sending area are limited to 1,000 square feet of floor area. Similarly, in the Cambria program of San Luis Obispo County, California, smaller lots are limited to a maximum of 900 square feet of total floor area unless they buy more floor area via the TDR program. In Miami-Dade County, Florida, developers who choose to build in the East Everglades sending area cannot let construction impede that sheet flow of water. The TDR program in San Francisco, California, has succeeded in permanently protecting 112 landmarks in part because it is practically impossible to demolish a structure designated as historically significant in San Francisco's downtown.

Off-site Requirements – Cities commonly require developers to pay for the infrastructure necessitated by new development either as actual off-site construction or monetary mitigation in the form of impact fees that the community then applies to transportation, sewer, water, and other public service improvements. Development in rural areas may not have the same level of cost for off-site improvements, perhaps because governments in rural areas may not appreciate the impact of cumulative development until they find traffic jams on their two-lane roads and human waste in local lakes and streams. When a jurisdiction requires greater off-site improvements in the receiving area than the sending area, it can create the unintended result of incentivizing development in sending areas (where off-site requirements are low). To avoid giving sending area development a competitive advantage over development in receiving areas, jurisdictions should be able to calculate the cost needed to provide adequate public services to each new development even if it's only an individual house. Rather than a uniform fee, the calculation should ideally account for the location of the proposed development. If a development is far removed from existing infrastructure, the calculation of development fees should incorporate the true cost of extending and maintaining roads, water mains, and other infrastructure to sparse development in these distant sending areas.

Success Factor Four - Few or No Alternatives to TDR

Many TDR programs underperform because developers have other ways of achieving additional development potential that do not involve TDR including competing incentives, alternative approval procedures, and exceptions to TDR.

Competing Incentives - Some jurisdictions allow bonus density in return for various amenities that are cheaper and/or easier for developers to provide. As detailed in the Part III profile, Portland, Oregon realized in a 2007 study that its transfer programs were in competition with18 bonus options. Some of these bonus options, particularly on-site locker rooms and eco roofs, were a cheaper means of gaining additional development potential than TDR. Not surprisingly, the locker room and eco-roof were more frequently used than the other options.

Even when competing bonus options have costs that are comparable with TDR, developers may prefer to add amenities to their own projects rather than provide a community benefit at an offsite location for the simple reason that on-site features increase development value while the value of offsite benefits gained by TDR may be difficult to market to prospective buyers. At an extreme, some jurisdictions grant bonus density in return for site design and architectural features that that are so fundamental that developments are likely to incorporate them with or without any bonus development incentives. In Birmingham, Michigan, TDR competes with four other ways of gaining bonus density, some of which developers might have planned to include regardless of incentives such as LEED certification, which adds value, improves occupancy, and reduces building operating expenses.

Conversely, successful TDR programs are likely to have few or no methods of achieving additional density other than TDR. In downtown San Francisco, California, where the TDR program has preserved 112 historic landmarks to date, TDR is the only way to exceed baseline development except affordable housing.

Alternative Approval Procedures – Some TDR programs have lackluster performance records because their jurisdictions offer alternative approval processes that allow bonus development potential at lower cost and/or with greater ease than TDR.

Cluster Subdivisions – Some cluster subdivision ordinances allow developers relief from minimum lot size requirements which can have the effect of allowing increased on-site density. For e xample, assume a 160-acre farm in a jurisdiction with a minimum lot size of 20 acres. This farm has a theoretical maximum density of one unit per 20 acres. But if half of this farm's acreage is not adjacent to a public roadway, the combination of minimum lot size and minimum lot dimensions could render it uneconomical to develop the interior of the property, perhaps reducing effective density to one unit per 40 acres. A cluster subdivision ordinance might allow eight lots to be created in a subdivision in which the jurisdiction allows smaller lots to be clustered on or near an existing public road in return for placing the eighth lot under a permanent agricultural easement. In this hypothetical example, the developer would be able to use the cluster option to economically increase on-site density to the theoretical maximum of one unit per 20 acres. Actually, many cluster subdivision ordinances grant bonuses that exceed theoretical as-of-right density limits in return for placing a minimum percent of the property under permanent easement. For example, Boulder County, Colorado, allows a baseline density of one unit per 35 acres to double when development is clustered on 25 percent of the property and a conservation easement is placed on 75 percent or more of the property.

Clustering can place a significant amount of land under easement. For example, the Boulder County program mentioned above resulted in 10,000 acres under easement in its first 15 years of operation. However, it is not always clear that cluster subdivisions are the best way of achieving community goals for preservation of agricultural activity and climate action. Clustering places residential development adjacent to land ideally destined for sustainable farming. But these residents may object to the dust, noise, odor, pesticides, and fertilizers needed for many types of commercial farming. As a result, the remnant parcels placed under easement may be confined to neighbor-friendly agricultural uses or perhaps not farmed at all. Also, clustering makes it economically feasible to locate small subdivisions across a formerly rural landscape. In some communities, this scattered development pattern may be an acceptable outcome. But, dispersed cluster subdivisions often require the wasteful extension of infrastructure and force their inhabitants to drive GHG-emitting cars to reach schools, stores, and everyday activities as well as job sites.

When a jurisdiction offers clustering as well as TDR, many developers choose clustering for several reasons. Unlike TDR, clustering reduces the number of participants needed for a successful outcome. In clustering, only one property is involved. In contrast, the developer of a TDR receiving site must find, negotiate, and buy TDRs to gain density (although there are many ways to reduce this complexity such as TDR brokers, banks, and allowing density transfer charges in lieu of actual TDRs.) Whether true or not, the perceived complexity of TDR transactions can cause some developers to choose the clustering alternative to save time and costs. In addition, development regulations and off-site requirements may favor cluster subdivisions over TDR receiving sites in urban areas, placing TDR at a competitive disadvantage as discussed above.

The TDR program in Brevard County, Florida underperforms in part because developers can use clustering to achieve the maximum densities allowed by the comprehensive plan, Similarly, when studying why its TDR program was languishing, Bainbridge Island, Washington, concluded that developers favored getting their additional density by clustering and by using some of the other options that generated additional development potential at lower cost.

Consequently, jurisdictions should consider whether clustering is actually implementing their goals for preservation, efficient delivery of public services, and climate action. Even if clustering is allowed, a government can take steps to try to make TDR at least as appealing if not more desirable. Boulder County, Colorado, for example, motivates property owners to transfer development potential to more appropriate receiving sites with a transfer of development rights option known as non-contiguous non-urban planned unit development, or NCNUPUD.

This mechanism can triple the density permitted by NUPUD (the county's name for a cluster subdivision). For example, a 35-acre parcel is allowed one unit as a matter of right, two units when the NUPUD option is used, or six units when transferred to a receiving site via NCNUPUD.

In addition to cluster subdivisions, planned unit developments, or PUDs, have a competitive advantage over TDR when jurisdictions fail to require TDRs for the additional development potential granted via PUD. Developers will logically avoid the cost and complexity of TDR if they can circumvent TDR requirements simply by pursuing a PUD mechanism that has no requirements for offsite preservation.

In some cities, zoning lot mergers (ZLM) outcompete TDR options. In a zoning lot merger, abutting parcels are treated as one for the purpose of calculating maximum allowable density. As described in Part III, between 2003 and 2011 385 transfers in New York City used ZLM while only two transfers occurred under New York's landmarks TDR code section (74-79) because the ZLM process is ministerial and therefore faster, cheaper and more certain than 74-79. The downside to ZLM in New York City is that it does not require the provision of community benefits. The 74-79 process, although disfavored by developers, nevertheless generated transfers from 12 landmarks between 1968 and 2013 (including Grand Central Terminal, Rockefeller Center, the Tiffany Building, and the Seagram Building), for a total transfer of 1,994,137 square feet.

Exceptions – Even more obvious, some TDR programs fail when their jurisdictions approve upzonings without requiring TDRs. These business-as-usual upzonings put TDR at a competitive disadvantage even if elected officials only grant them to developments that are considered exceptional. When exceptions to TDR are granted, it becomes harder to tell the next developer that he or she has to use TDR when several predecessors were able to avoid it simply by asking for rezonings.

As a case in point, the January 24, 2015 edition of SRQ Daily reported that the Board of Commissioners in Sarasota County, Florida had granted permission for more than 5,000 additional homes in land use changes that circumvented TDR when the developers claimed TDR would be an unreasonable expense.

In contrast, successful TDR programs are not likely to let developers use zone map amendments to avoid TDR. In Howard County, Maryland, where TDR has protected more than 4,900 acres, it is nearly impossible to circumvent TDR by using rezoning. Similarly, Summit County, Colorado, where TDR has protected over 2,500 acres to date, attributes much of its success to precluding upzonings without TDR.

Some jurisdictions in fact tie upzonings directly to their TDR programs. In Livermore, California, general plan amendments that have the effect of increasing residential density automatically become TDR receiving sites in which the density allowed by the former zoning becomes baseline density and all development in excess of baseline is subject to TDR compliance requirements.

The success of the TDR program in Summit County, Colorado, which has preserved over 2,500 acres to date, can partly be attributed to the fact that the county as well as the participating towns of Breckenridge and Blue River have policies prohibiting upzonings without the use of TDR.

Success Factor Five – Market Incentives

Many TDR programs allow one bonus dwelling unit on a receiving site for each dwelling unit precluded on a sending site, a so-called one-to-one transfer ratio. In some cases, this ratio supports a viable TDR market. However, in many cases, a one-to-one transfer ratio produces a mismatch in how sending area landowners and receiving site developers view the value of a TDR. For example, if a sending area allows on-site development at a density of one unit per acre and the TDR grants one TDR per one foregone dwelling unit, that TDR represents the value of placing 40 acres of sending area land under permanent easement. In many places, the value of a 40-acre conservation easement might be so high that a receiving area developer cannot afford it if the TDR allows only one bonus unit on a receiving site. Unfortunately, this value mismatch causes many programs to underperform.

Planners should find one or more ways of estimating the value of development potential gained and lost when they are developing a TDR program. To simplify this discussion, assume that a county wants to motivate the relocation of single-family residential unit potential from its Agricultural (A) Zone to its Suburban (S) zone. The S zone, the receiving area, allows two units per acre and development professionals agree that there is a market for higher density in this zone, that two units per acre could become baseline, and maximum density in the S zone could be five units per acre when developers retire one TDR for each bonus unit of density over baseline. Ideally, the county would hire an economist to estimate the average value increase in one or more typical, vacant parcels in the S zone for each unit allowed in excess of baseline density. Assume the economist estimates that each additional unit adds $40,000 of value to a subdivision in the two- to five-unit-per-acre density range. The planners do not want to remove developer motivation to exceed baseline so use a working number of $20,000 per TDR as a target, subject to policy decisions by appointed and elected county officials.

If the county in this hypothetical example chooses not to hire an economist, the planners can perform their own estimates ideally in cooperation with development professionals. In communities where development is brisk, developers and real estate agents may already know or can readily estimate how much more they can afford to pay for land on the basis of each additional dwelling unit that the parcel can accommodate. These estimates can be compared with the prices paid by developers of comparable receiving sites in jurisdictions with comparable real estate markets. Developers in Chesterfield Township, Burlington County, New Jersey, were willing and able to pay $50,000 per TDR, which represented roughly ten percent of the retail price of a new single-family home with a lot in the receiving area. In deciding whether or not Chesterfield is a good comparison, consider that Chesterfield is a well-preserved horse farm area situated between New York City and Philadelphia, where the development community had participated in the planning of the receiving area (a multi-use village with shops and a school), and where developers were able to build according to the plan without the delay, cost and uncertainty of a discretionary approval process. A TDR program can maximize developer motivation by observing most or all of the ten success factors. However, not all jurisdictions will have the locational advantages of Chesterfield which is why it is important to choose comparable communities when estimating how much developers will be willing to pay for TDRs.

Once the planners in this hypothetical example have estimated that developers should be able and willing to pay $20,000 for each additional single-family lot above baseline in receiving areas, they can similarly estimate how much sending area property owners will be willing to place under easement for a TDR that they can sell for $20,000. Again, ideally, the jurisdiction will hire an economist. But planners can also get a sense of sending area owner expectations from the per acre prices paid for conservation easements on comparable property by purchase of development rights (PDR) programs and other preservation initiatives that may operate in the area. In addition, planners may be able to estimate expectations by interviewing sending area property owners. It should be noted that value estimates do not appear in a TDR ordinance; these value estimates are simply background information used to make necessary adjustments to the TDR program components.

In the hypothetical county above, assume the planners know that easements on farmland are selling for an average of $2,000 per acre. Consequently, it seems unlikely that a TDR program will work at a one-to-one transfer ratio since sending area owners are estimated to want $80,000 to place an easement on 40 acres at $2,000 per acre. The most commonly used solution to this kind of mismatch is to adopt a TDR allocation ratio, meaning the number of acres needed per TDR in order to form a TDR value that is attractive to both buyers and sellers. In this hypothetical county, an allocation ratio of one TDR per ten acres would tend to create a TDR valued at roughly $20,000, the value that should satisfy sending area owners' desire for $2,000 per acre while creating a $20,000 TDR that developers are estimated to be able and willing to pay (assuming each TDR allows one bonus dwelling unit above baseline as discussed above.) Note that the allocation rate differs from the density that the sending area owner can build on site, a difference that is commonly found in successful TDR programs.

However, there may be some reason why this hypothetical county wants on-site development potential and the allocation ratio to be identical. If so, the program could address the value imbalance by adjusting the allowance ratio, meaning the amount of additional development potential allowed per TDR. In our hypothetical, the program could allow receiving area developers four bonus single family lots for each TDR acquired from a sending site. This adjustment of the allowance ratio achieves that same result as the allocation ratio of one TDR per ten acres discussed above: the sending area property owners are paid $2,000 per acre, or $80,000, for placing 40 acres under easement and the receiving site developers are allowed to build four units above baseline using one $80,000 TDR that permits four bonus units, at the cost of $20,000 per bonus unit that these developers are estimated to be able and willing to pay.

Adjustments in the allocation of TDRs to sending areas and the per-TDR allowance ratios to receiving sites are indications that a jurisdiction understands the importance of using program components to create a functioning market. A 2009 study of the TDR programs that had saved the most land (at that time) revealed that 15 out of 20 had observed this success factor.

TDR programs that adjust these components based on economic considerations often see improved performance. As detailed in the profile in Part III, the TDR program in Douglas County, Nevada languished originally, largely because it granted only two TDRs per 19 acres of sending area land placed under easement, an allocation ratio that made TDRs too expensive for receiving site developers to afford. In 2001, Douglas County increased the allocation ratio to nine additional TDRs per 19 acres plus extra TDRs for preservation of floodplains, rights to irrigation water, and public access easements to bodies of water, public lands, or historic sites. Using all of these options, a sending site owner can now transfer up to 25 TDRs per 19 acres. By 2009, 4,003 acres had been placed under easement including 2,892 acres in floodplains, an indication that the allocation formula is implementing the goal of Douglas County's Living River policy of retaining rivers in their natural state and allowing water to access the floodplain.

Similarly, in 2007, Marion County, Florida changed its original allocation formula from three TDRs per ten acres to one TDR per preserved acre for sites at least 30 acres in size. This change energized the market, and the program has now preserved at least 3,580 acres.

Establishing a TDR allocation ratio is easier when land throughout the sending area has relatively similar development value. However, larger sending areas often have diverse value. Some TDR programs apply a uniform allocation ratio despite variations in development potential and value. For example, Montgomery County, Maryland, successfully offers one TDR per five acres of protected land across its 93,000-acre Agricultural Reserve. Despite this large sending area, Montgomery County's TDR program protected over 52,000 acres in places close to existing development and infrastructure as well as at the periphery of the Ag Reserve, possibly because the county committed to preserving the entire reserve by not rezoning a single acre of sending area land after the TDR program was adopted in 1980.

Other jurisdictions aim for comparable levels of preservation throughout a diverse sending area by using various allocation ratios representing different categories of development value and/or preservation priorities. Some jurisdictions use the difference in maximum zoning densities to serve as a surrogate for differences in development value when several zones exist within the sending area. For example, if a sending area encompasses two zones, some jurisdictions might have one TDR allocation formula for the sending area that allows lower density to on-site development and a higher ratio to land in a zone that allows higher density to on-site development. As noted above, Douglas County, Nevada, allocates extra TDRs for sending sites that provide high-priority public benefits like floodplain preservation and public access. In the New Jersey Pinelands, the number of PDCs available to a sending site is based on both development potential and environmental sensitivity. For example, the Preservation District allocates one PDC per 39 acres of uplands, 0.2 PDCs per 39 acres of wetlands, two PDCs per 39 acres of land approved for mining but undisturbed, but no PDCs to land already mined.

Some programs go to extraordinary lengths to minimize the ability of developers to save money by obtaining TDRs from sending sites with the lowest development value. The owners of land in the sending area of San Luis Obispo's countywide TDR program wanted an allocation formula that created a level playing field giving all owners a roughly equal chance of selling their TDRs. Under one of two allocation methods, sending site owners interested in participating have the development value of their land appraised. The owners receive one TDR for each $20,000 of development value calculated by these site-specific appraisals. Holders of TDCs sell them at prices negotiated between buyers and sellers; the formula in which appraised sending site development value is divided by $20,000 only determines the number of TDCs allocated to a sending site. Plymouth, Massachusetts also uses site-specific appraisals to determine sending area TDR allocations.

The Rural Land Stewardship (RLS) Program in Collier County, Florida developed unique allocation and allowance formulas that have succeeded in preserving 54,962 acres to date. First, the land itself is assigned preservation value according to the priority status of its environmental resources, meaning habitat, flow-ways, or aquifer recharge zones. In the second step, the applicable property owners choose the extent to which they want to restrict the use of their property. In tiers of restrictions, owners can decide to only prohibit residential development or also restrict other activities such as mining, recreation, cropland, agricultural support, and pasturing. The amount of credits available to a property grows as the owner chooses to apply greater restrictions. In Ave Maria, the receiving area of the RLS program, eight TDRs are required to develop each acre of receiving site land, an approach that maximizes the potential to achieve the densities called for in the receiving area plan.

TDR program receiving areas can also have many sub-areas, each with different baselines and various types of additional development potential, ideally based on market considerations to achieve transactions that are mutually beneficial for TDR buyers and sellers. In addition to multiple receiving areas, each with their own components, TDR programs may offer different types of bonus development potential in different receiving areas. Ideally, each conversion factor from one type of development potential to another is based on an economic analysis to, again, create transfers that appeal to sending area landowners and receiving area developers. For example, in the Miami-Dade County, Florida Severable Use Right (SUR) Program, receiving areas are properties with residential, commercial, and industrial zoning designations located within Miami-Dade County's Urban Development Boundary (UDB). All Miami-Dade County zones designated for urban development can receive SURs with the exception of the agricultural, environmental, recreation, and open space zones. In four residential zoning districts, developers can use SURs to reduce minimum lot size and minimum frontage. In four additional residential zoning districts, SURs can be used to reduce maximum lot coverage as well as minimum lot size and frontage. In three more residential zones, SURs can increase maximum height limits as well as density, floor area ratio, and coverage. In two additional residential districts, SURs can increase maximum height, density and floor area ratio but not coverage. In the PAD, ECPAD, and REDPAD districts, SURs can increase density by 20 percent. In the Core or Center Sub Districts of Community Urban Center zoning districts with certain designations, baseline density can be increased by up to eight units at the rate of two units per SUR. In seven commercial districts, baseline FAR can be exceeded at the ratio of 0.015 FAR per SUR in seven commercial districts and by 0.010 FAR in the OPD district.

In Pierce County, Washington, the number of bonus dwelling units allowed by TDR is calculated by subtracting the original (pre-transfer) units from the final (post transfer) units and dividing by a conversion factor specified for each of 25 different final density categories ranging from one unit per acre to 25 units per acre. The conversion rate increases as the final density increases. For example, a one-acre site with an original density of four units per acre and a final density of 14 units per acre would require two TDRs because the difference of ten bonus units is divided by five, the conversion rate for the 14-unit per acre final density category. The conversion rate tops out at eight for projects within the 25-unit per acre density category.

Success Factor Six – Certainty of Ability to Use TDR

Approval procedures that involve public hearings and discretionary decisions can create delays, added costs, and even project denials. In contrast, administrative approval procedures give developers confidence that they can plan and schedule TDR receiving site projects with a fair degree of certainty of staying on schedule and within budget.

As discussed in Parts I and III, the Rural Land Stewardship TDR program in Collier County, Florida assured developers that they could build in accordance with the master plan for the Ave Maria receiving area if they complied with all regulations and retired the required number of TDRs. Similarly, developers in the Old York Village receiving area of Chesterfield Township, Burlington County, New Jersey, knew that they could get administrative approval by following the rules for that program. Developers were engaged in the creation of these two TDR programs, and it is noteworthy that administrative approval became a central element in both.

Lumberton Township, Burlington County, New Jersey, also offered administrative approval to a TDR program that succeeded in preserving 850 acres of farmland. Lumberton went to great lengths to ensure that its goals for the receiving area would be implemented without the need for public hearings or discretionary decisions on each individual development. The receiving area for Lumberton's first TDR program included 508 acres of land mapped as receiving area in the township master plan. Significantly, the use of TDR was granted administratively as long as the receiving site project adhered to all development requirements, including 35 pages of standards and guidelines found within the TDR code section itself addressing the retention of natural elements and cultural features, as well as the provision of storm water management facilities, public utilities, and landscaping, Since the receiving area was adjacent to the Historic District of Lumberton Village, the TDR code also required new buildings to be compatible in scale with existing structures and reflect the architectural styles of the 18th, 19th and early 20th centuries including the Colonial, Georgian, Federal, Greek Revival, and Victorian styles. To emphasize this requirement, the TDR code incorporated 17 pages of architectural standards regarding façade treatment, building materials, fenestration, rooflines, fences, and open space. While this level of detail is unusual in a TDR code, it may also explain why Lumberton was able to approve receiving site projects without the discretionary review that can cause the delay, redesign, and added costs that sometimes discourage developers from choosing the TDR option.

The TDR program in Howard County, Maryland, which had preserved 4,980 acres of farmland as of 2016, also allows the use TDRs as a matter of right. Other TDR programs that offer administrative approval of TDR transactions include: Greenville, South Carolina; Lee County, Florida; and Stafford County, Virginia.

Of the 20 programs examined in the 2009 study, 14 were considered to provide a reasonable level of certainty that developers could proceed with TDR receiving site projects if they complied with all requirements: King County, Washington; New Jersey, Pinelands; Palm Beach County, Florida; Collier County, Florida; Calvert County, Maryland; Queen Anne's County, Maryland; Sarasota County, Florida; Pitkin County, Colorado; Boulder County, Colorado; Howard County, Maryland; Miami-Dade County, Florida; Charles County, Maryland; and Chesterfield Township, Burlington County, New Jersey. However, the fact that six of these 20 programs did not offer certainty suggest that this should be considered as a helpful but not necessarily essential success factor.

Success Factor Seven – Strong Public Support

The most successful jurisdictions exhibit strong public support for the goals of their TDR programs. Of the 20 top programs explored in 2009, 13 combined TDR with other preservation tools, including new voter-approved tax measures dedicated to conserving farmland, environmental areas, and other priority resources. Although a big advantage of TDR is its reliance on private sector funding, the coexistence of tax-funded tools suggests that a majority of the population understands the importance of preserving its green infrastructure. TDR programs must be able to survive from one election to the next. New public officials could be convinced to allow exceptions to TDR requirements or even abandon programs entirely unless the general public maintains its support for these programs.

As described in Parts I and III, the TDR program in Montgomery County, Maryland, protected over 52,000 acres of the 93,000-acre Agricultural Reserve. However, Montgomery County saved another 20,000-plus acre of the Ag Reserve with a conservation easement donation program, two state-funded purchase of development rights programs, and the Montgomery County Agricultural Easement Program that preserves land with county tax revenues. Given this level of commitment to conservation, it was not surprising that Montgomery County has approved one of the most ambitious climate action plans in the country with a goal of eliminating all GHG emissions by 2035. Part of that plan relies on the Ag Reserve to maintain compact urban form, protect forestry, watersheds, and ecosystem resources, as well as maintain an ample amount of farmland for restorative agriculture.

King County, Washington, leads the nation with 144,500 acres protected by TDR alone. But people here are also willing to contribute their own money to open space protection as seen by the fact that voters have repeatedly approved multi-million-dollar conservation bonds and annually dedicate roughly $10 million of property tax revenue to their Conservation Futures Fund. Instead of using its publicly generated funds to buy only land or easements, King County sometimes applies this money to the acquisition of TDRs severed from selected sending sites. By banking and selling these TDRs, King County effectively transforms what would otherwise be a one-time purchase into an ongoing revolving fund for preservation.

Similarly, TDR is just one of several preservation strategies used by Boulder County, Colorado to preserve a 94,000-acre open space and trails network in a county where more than two-thirds of the total land area is saved within national parks, forests, and other public spaces. TDR was one of many tools used to form a 100,000-acre greenbelt credited with protecting the City of Boulder from wildfires as well as safeguarding natural resources, biodiversity, and opportunities for outdoor recreation.

Other notable examples of jurisdictions where strong public support for preservation promotes this success factor for TDR programs are profiled in Part III. The State of New Jersey reflected public support for preservation of the New Jersey Pinelands TDR program by capitalizing the Pinelands Development Credit Bank, approving a $50 million infrastructure program for receiving areas, and launching a separate program to buy and retire credits. The TDR program in Palm Beach County, Florida was adopted after the voters had approved $250 million in environmental preservation bonds. In Collier County, Florida, where three separate TDR programs have preserved over 62,000 acres, more than 80 percent of the county is protected by a combination of federal, state, county, and private efforts. Suffolk County, New York has spent more than $1 billion on preservation programs, an indication of strong public support for the protection offered by TDR.

Success Factor Eight – Program Simplicity

Program simplicity is a helpful characteristic because the mechanism should ideally be understandable to the multiple stakeholders involved including landowners, developers, homeowners groups, preservationists, appointed/elected officials, and the general public. In the 2009 paper, 12 of the 20 studied programs were considered to be relatively simple, suggesting that this factor can be beneficial although not necessarily essential to TDR program success.

The success of the Montgomery County TDR program is partly due to its simplicity. The entire 93,000-acre sending area has a single zoning designation with one on-site maximum density, and one consistent allocation ratio. Although several zones can receive TDRs, the per-TDR allowance ratio is uniform in each zone. As an indication of its simplicity, the Montgomery County TDR program has preserved over 52,000 acres without the need for a TDR bank.

Some programs are initially complicated but look for ways over time to simplify. Until recently, the

TDR program of the Tahoe Regional Planning Agency (TRPA), California/Nevada was complicated by a requirement that interjurisdictional transfers had to be approved by both jurisdictions and by limited ability to convert three types of marketable commodities. These and other complications were slowing down the replacement of older, environmentally harmful development with new construction designed to minimize environmental impacts. In 2018, TRPA simplified the program by eliminating the interjurisdictional approval requirement and adopting ratios allowing conversions between residential units of use, commercial floor area, and tourist accommodation units.

In addition to adopting simple regulations, at least 31 TDR programs simplify transactions by allowing receiving area developers to make a monetary payment known as a density transfer charge (DTC) instead of presenting actual TDRs for extinguishment. For each additional residential unit allowed through a rezoning in Berthoud, Colorado prior to 2009, developers could chose to either preserve one acre of land or pay a DTC of $3,000 per additional single family residential unit or $1,500 per extra multiple-family residential dwelling unit. The town dedicated 94 percent of DTC revenues to the preservation of land and allowed the remaining six percent to cover administrative costs.

In Gunnison County, Colorado, developers can elect to become receiving sites and increase density within parameters by paying a DTC calculated as ten percent of the site's land value generated by the increased development potential. This program and five others do not offer developers the option of using actual TDRs; DTC is only method of compliance. In Gunnison County, developers can defer some or all of the required payment upon final plat recordation. The county adds these payments to its Land Preservation Fund, which uses open space sales tax revenue and state grants as well as DTC proceeds to fund the winners of an annual competition of proposals submitted by property owners wanting to preserve their land. As a result, Gunnison County's approach simplifies program adoption, administration, and compliance. Sending areas are not designated and a TDR allocation formula is not contained in its TDR ordinance, thereby avoiding disputes that sometimes arise over who can sell TDRs and how many TDRs will be issued in return for preserving a sending site. Essentially the sending site is selected by the county when it decides where to spend Land Preservation Fund dollars and the amount of compensation is proposed by the owners who want to protect their properties. Similarly, Gunnison County did not need an economic study to determine how many TDRs would be needed per unit of bonus density because TDRs are not involved. Furthermore, Gunnison County developers do not have to negotiate and buy TDRs from sending site owners because there are no TDRs, and the county performs the sending site preservation. In addition, because compliance is in the form of a cash payment, the county is relieved of having to create, track and retire actual TDRs. Using cash also allows the county to target the best acquisitions and gain maximum leverage from DTC revenues.

While a DTC-only process streamlines administration, in most programs, DTC is offered as an option that developers are free to use or decline in favor of transacting actual TDRs. The experience of Charlotte County, Florida, suggests that developers appreciate this ability to choose. In Charlotte County, the builders of small projects chose to comply with DTC, possibly because it is not economical for them to find, negotiate, and buy actual TDRs. Conversely, the developers of large subdivisions prefer to comply with actual TDRs, likely because this is less expensive for them due to economies of scale.

Jurisdictions that allow developers the choice of complying with DTC or actual TDRs include: Austin, Texas; Berthoud, Colorado; Charlotte County, Florida; Clifton Park, New York; Easthampton, Massachusetts; Goshen, New York; Hadley, Massachusetts; Larimer County, Colorado; Livermore,

California; Los Angeles, California; Los Angeles County, California (Malibu Coastal Zone); Oxnard, California; Palmer, Massachusetts; Pierce County, Washington; Red Hook Township, Dutchess County, New York; San Diego, California; Scarborough, Maine; St. Mary's County, Maryland; Summit County, Colorado; Summit County, Utah; Tacoma, Washington; West Valley, Utah; Westfield, Massachusetts; and Whatcom, County, Washington. Programs that use DTC exclusively include: Bellingham, Washington; Fort Myers, Florida; Gorham, Maine; Gunnison County, Colorado; Hatfield, Massachusetts; Sussex County, Delaware; and Warwick, New York.

Success Factor Nine – Promotion and Facilitation

TDR programs are more likely to be effective when planners make the effort to promote and facilitate them. Educational materials explaining the TDR mechanism and its benefits remind potential stakeholders of this option and also indicate that a jurisdiction is serious about putting TDR to work.

Montgomery County, Maryland regularly demonstrated its commitment to TDR with detailed studies, status reports, and informational brochures such as *Plowing New Ground*, a user-friendly guide that clearly spelled out the need for TDR, the advantages for landowners and developers, plus step-by-step procedures for navigating transactions from start to finish.

King County, Washington organizes everything you ever wanted to know about their TDR program onto its web site with links for landowners and developers as well as information on the TDR market, TDR Bank, TDR code, TDR incentives, and the TDR Exchange where buyers and sellers can find each other.

The website for the New Jersey Pinelands Development Credit Program also gathers studies, forms, transaction information, and sales activity in a single location. Furthermore, the Pinelands Commission routinely sponsors educational programs, hikes, and other activities that encourage the general public to appreciate this unique ecosystem and ideally support its continued protection.

In addition to TDR banks, which are discussed in the next section, jurisdictions can facilitate TDR programs with various incentives and accommodations. Many of these features were discussed above such as the funding of receiving area infrastructure provided in the New Jersey Pinelands program and the ability to defer DTC payment to the sale of a receiving site home offered in Gunnison County, Colorado.

The TDR program in Warwick Township, Lancaster County, Pennsylvania demonstrates how government facilitation can generate success. In most PDR programs, public funds are used to purchase easements rather than TDRs, resulting in the need to generate more public funding in order to preserve more land. Warwick partners with the Lancaster Farmland Trust, a private, non-profit conservancy, and the Lancaster County Agricultural Preserve Board to buy TDRs. Even though some of the funding for these purchases is from the county PDR program, the township and the trust retain and sell these TDRs with the understanding that all sale proceeds are used exclusively to buy TDRs from Warwick sending sites. In effect, this process turns what would otherwise be a single acquisition into a perpetual farmland preservation revolving fund.

Notably, eight of the 20 TDR programs examined in the 2009 article were not found to excel at promotion or facilitation, emphasizing that these characteristics are helpful but not necessarily essential to creating a successful TDR program.

Success Factor Ten – TDR Banks

A TDR bank is a public or quasi-public agency authorized by a jurisdiction primarily to buy, hold, and sell TDRs. The bank can counteract economic cycles by buying TDRs during construction downturns and selling them when demand returns. Banks can provide relief to sending area property owners having trouble finding a buyer for their TDRs. Banks can simplify compliance for receiving area developers by providing a source of readily available TDRs, thereby avoiding the time and cost of having to find, negotiate, and buy TDRs from individual sending site owners. Banks can jumpstart transactions at a time when the players in the private market are waiting for others to make the first move. Banks can help set and stabilize TDR prices. Banks can convert traditional one-time acquisitions of land and easements into ongoing revolving funds in which the revenue generated by the sale of past preservation is used to pay for the TDRs representing future preservation.

The King County's TDR Bank is a big reason why this program leads the nation in protected land, with 144,500 acres preserved to date by TDR. The TDR Bank here is tasked with three roles: 1) Facilitating private transactions by buying TDRs when sending site owners want to sell and holding them until receiving site developers want to buy them; 2) Acting as a revolving fund by reinvesting the proceeds of its TDR sales in continuing land protection; and 3) Acquiring sending sites of special significance to incorporated cities as a means of motivating these cities to enter into interlocal TDR agreements with King County.

In addition to these three primary tasks, the Bank sells options to buy TDRs, allowing developers to secure a specific number of TDRs at a known price for the period of time needed until receiving site project approval. The Bank also offers to create extended purchase and sale agreements allowing developers to reduce risk and up-front costs. The Bank incentivizes cities to enter into interlocal TDR agreements with King County because the Bank can negotiate revenue sharing provisions that dedicate a portion of the proceeds from TDR Bank sales to the funding of infrastructure, parks, streetscapes and other amenities within the participating city's receiving area.

King County's TDR Bank sometime buys TDRs using its Conservation Future Tax. This tax is made possible by a State of Washington law allowing counties to dedicate a portion of property tax exclusively to preservation projects. The King County Conservation Futures Tax typically generates about $10 million annually for conservation projects. By using Conservation Futures Tax revenue to buy TDRs, King County converts what would otherwise be a one-time acquisition into an ongoing revolving fund for preservation. In its most dramatic example, King County used $22 million of Conservation Futures Tax proceeds to purchase 990 TDRs representing the preservation of 90,000-acre forest 25 miles east of Seattle.

As a disadvantage, TDR banks must be stocked with TDRs, a process that typically requires public funding, which can be opposed by taxpayers. However, several communities use TDR banks to leverage tax revenues that would occur with or without a TDR program and use that leveraged money to accelerate preservation. Palm Beach County, Florida passed $250 million in conservation bond measures used to acquire its 31,000-acre nature preserve system. By severing, banking, and selling these TDRs, Palm Beach County uses TDR sale proceeds to expand and maintain these preserves. Likewise, when King County, Washington, uses part of its property tax revenue to buy and bank TDRs, it can use revenue from the sale of those TDRs to accomplish further preservation and incentivize incorporated cities to establish TDR receiving sites for inter-jurisdictional transfers.

Palo Alto, California, sells TDRs from its city-owned landmarks and uses the proceeds to fund the rehabilitating of these public buildings. Other jurisdictions that treat the development potential of their properties like banks include: Farmington, Utah; Issaquah, Washington; Largo, Florida; Los Angeles, California; Madison, Georgia; Morgan Hill, California; and Sunny Isles Beach, Florida.

Banking has been essential to the success of TDR in Seattle, Washington. Seattle's TDR bank has bought and sold floor area representing the preservation of affordable housing, landmarks, and landmark performing arts centers as well as the development of its concert hall and sculpture park. By 1998, the TDR Bank had purchased 274,340 square feet of floor area from eight buildings, including the Paramount Theater and the Eagles Auditorium, and had sold 249,380 square feet. In addition, the city had 423,000 square feet of Major Performing Arts Facility TDRs to sell, with the sale proceeds to be used to pay the debt service on construction of Benaroya Symphony Hall. In a single transfer to the Washington Mutual office tower in 2004, the TDR bank sold 9,842 square feet of Major Performing Arts Facility TDRs that generated $150,000 in debt payment for Benaroya Hall, 90,728 square feet of open space TDR putting $1.3 million toward the Seattle Art Museum's Olympic Sculpture Park, and 132,500 square feet of Housing TDR directed at the preservation of downtown affordable housing.

As detailed in Chapter 7 on wildfire adaptation in Part I, the California Coastal Conservancy and the Mountains Restoration Trust created banks that were instrumental in the success of the Malibu Coastal Zone TDR program in Los Angeles County, California, during its heyday. In addition to significant sending site acquisitions, these entities creatively combined donated with purchased TDRs to offer TDRs to receiving site developers at prices they were able and willing to afford.

While most TDR banks are public agencies, the Land Conservancy of San Luis Obispo County, a private, non-profit organization was authorized to administer banking functions for the Cambria TDR program in San Luis Obispo County, California. The Land Conservancy bought the first sending area easements with a $275,000 loan from the California Coastal Conservancy and proceeded to sell transferable square footage at prices that covered the cost of administration plus the acquisition of 350 parcels of land and the conservation of the Fern Canyon Preserve, home of the rare Monterey pine. The Land Conservancy demonstrates the possibility of using land trusts as TDR bank administrators plus the way in which a TDR bank can use a modest amount of seed money to maintain a continuously functioning TDR program.

As further described in Part III, the TDR programs with banks include: Central Pine Barrens, New York; Issaquah, Washington; King County, Washington; Los Angeles County, California; Manheim Township, Lancaster County, Pennsylvania; Morgan Hill, California; New Jersey Highlands, New Jersey; New Jersey Pinelands, New Jersey; New York City, New York; Palm Beach County, Florida; Pierce County, Washington; San Luis Obispo County, California; Seattle, Washington; South Lake Tahoe, California; Southampton Town Suffolk County, New York; Southold Town, Suffolk County, New York; Summit County, Colorado; Tahoe Regional Planning Agency, California/Nevada; Warwick Township, Lancaster County, Pennsylvania; and Woolwich Township, Gloucester County, New Jersey.

Only four of the top 20 US TDR programs in 2009 had TDR banks although these four programs had preserved a large amount of land. The other 16 TDR programs succeed without the benefit of a bank, including the highly regarded program in Montgomery County, Maryland. Consequently, banks are considered helpful but not necessarily essential to the success of a TDR program.

References

McConnell, V., M. Walls, and F. Kelly. 2007. Making TDR Programs Better: Report Prepared for the Maryland Center for Agroecology. Accessed 1-29-21 at Microsoft Word - McConnell Walls.FINAL-2.doc (umd.edu).

Pruetz, R. and N. Standridge. 2009. What Makes Transfer of Development Rights Work? Success Factors from Research and Practice. Journal of the American Planning Association. Vol. 75, No. 1, Winter 2009.

Walls, Margaret & Virginia McConnell. 2007. Transfer of Development Rights in U.S. Communities. Washington, D.C.: Resources for the Future.

Montgomery County, Maryland observes almost all of the ten TDR success factors.

PART III

Profiles of US TDR Programs

Part III provides profiles of 282 TDR programs in the United States. In many cases, more detailed versions of shorter profiles can be found under the TDR Updates tab at the website www.SmartPreservation.net.

Acton, Massachusetts, population 23,662 (2019), is located in Middlesex County, 21 miles west/northwest of Boston. The TDR bylaw here allows the transfer of residential and non-residential development potential from a highway corridor to mixed-use village centers in order to promote compact development in areas more efficiently served by infrastructure. Residential development potential cannot be converted to non-residential floor area but non-residential sending site potential can be converted to residential receiving site development at the ratio of one bonus dwelling unit per 1,000 square feet of non-residential sending site floor area. Receiving site developments that use the TDR option must be mixed use projects. In 2016, the Town approved a mixed-use TDR receiving-site project in East Acton Village incorporating 8,000 square feet of commercial space and 14 dwelling units.

Adams County, Colorado, population 517,421 (2019), includes some of the northern suburbs of Denver, extends east for 70 miles into the plains, and wraps around Denver International Airport. Section 3-28-03-06 of the Adams County Development Standards and Regulations, dated 2005, allows transfers of development rights through the planned unit development (PUD) process to provide economic relief and alternatives to landowners subject to development restrictions within floodplains, habitat areas, significant agricultural lands and areas affected by the Denver and Front Range airports. TDRs are allocated based on the underlying zoning of the sending areas, which typically means one

TDR per 35 acres. Adams County then applies the following transfer ratios based on the nature of the sending area.

1) Barr Lake/South Platte River - 25:1 (meaning 25 bonus units can be built at an approved receiving site for each TDR transferred from this type of sending area based on a conservation easement precluding the development of 35 acres of that sending site);
2) Natural Resource Conservation Overlay – 15:1;
3) Important Farmlands – 10:1;
4) Airport Influence Zone – 5:1.

Six receiving areas are identified geographically. As of 2015, from 3,000 to 4,000 acres were estimated to be preserved by TDR.

Agoura Hills, California, population 20,222 (2019), is located between the Santa Monica Mountains and the Simi Hills, 30 miles west of downtown Los Angeles. Much of the land within the current city limits had already been developed or approved for development by the time Agoura Hills was incorporated in 1982. To promote preservation of the remaining open space, the city zoned a considerable amount of hillside land as open space with a maximum density of one unit per five acres and additional restrictions based on steepness of slope. For example, a lot with a slope of more than 35 percent must provide 20 acres per dwelling unit. In 1987, the city provided mitigation with a transfer of development credits ordinance. The number of credits available for transfer is equal to the number of units allowed on the sending site, representing a one-to-one transfer ratio. The receiving sites must be in one of four residential zoning districts. Transferred development credits cannot be used to increase the density of the receiving site more than 20 percent higher than the density allowed by zoning for that site. Developers can also increase density for on-site open space preservation using the City's Cluster Development Ordinance as well as certain aesthetic enhancements and infrastructure improvements. As of 2005, roughly 700 acres had been saved through easement donations, but the TDR provisions had not experienced any transfers.

Alachua County, Florida, population 269,043 (2019), is located in northern Florida and surrounds the City of Gainesville. In 1987, the county adopted TDR provisions aimed at preserving the environmental, cultural, and historic character of Cross Creek, the village that was the setting for many of the writings of Marjorie Kinnan Rawlings. In this program, transfers are limited to adjoining properties, using a single, unified application for a Planned Development. The 2001-2020 Comprehensive Plan provides the following explanation of the sending site as-of-right development potential and transfer opportunities.

- In areas designated as wetlands, density transfers are allowed at a ratio of one unit per five acres to contiguous property.
- In Exceptional Upland Habitat, development can occur on site at a density of one unit per five acres but transfers are allowed at a ratio of two units per five acres to contiguous receiving sites.
- In Hammock Zone, on-site development is limited to one unit per two acres in the Village Center and one unit per five acres in the Village Periphery, but transfers can occur at the ratio of two units per five acres.
- In Lake Buffer Areas, density transfers are allowed at the ratio of two dwelling units per five acres.
- In Bald Eagle Nesting Areas, transfers to contiguous property are permitted at a rate of two

units per five acres in secondary zones and three units per five acres in primary zones.

Development rights may be transferred to receiving sites in the Village Periphery Development Area with a baseline density of one unit per five acres which can be increased to one unit per acre in order to accommodate transfers from the sending sites described above.

In 2009, Alachua County added a county-wide TDR program designed to protect environmental resources, viable agriculture and rural landscape. TDR allocation to sending sites is the lesser of the following two options minus any dwelling units not included in the transfer:

- The number of dwelling units allowed on the sending site; or
- The number of upland acres on the property.

In addition, two bonus TDRs are granted per site plus one TDR per ten acres of conservation area on site and one TDR per 20 acres of non-conservation area on site. Sending site easements may allow the sending site owners to retain residential density of one dwelling unit per 40 acres on agricultural land and one dwelling unit per 200 acres on conservation land (or as high as one unit per 40 acres under specified conditions such as the clustering of units to best protect environmental resources.)

The county-wide program has three types of receiving areas. 1) Non-residential development in the unincorporated area can reduce its on-site open space requirement by one acre for every 10 TDRs. 2) Any proposed amendment to expand the Urban Cluster must include a commitment to buy two TDRs per unit of proposed increase in residential density or ten TDRs per acre of non-residential land use created. 3) By interlocal agreement, receiving areas can be located within any municipality in Alachua County.

American Fork, Utah, population 33,161 (2019), surrounds the mouth of the American Fork River where it flows into the northeastern corner of Utah Lake 20 miles south of Salt Lake City. Municipal Code Section 17.4.605: Transferable Development Rights Overlay designates mapped sending and receiving areas. The number of TDRs available to a site in the TDR Sending Area Sub-Zone varies depending on whether the underlying zoning is the Shoreline Preservation (SP) Zone or some other zoning district. The SP Zone includes areas near Utah Lake that are inundated and/or subject to periodic flooding. This zone allows farming and golf courses but no residential development. For the purpose of TDR, the development rights can be created at the rate of one transferable unit per five acres in the SP zone. In all other zones, the number of development rights available for transfer is one-half of the base density, meaning the maximum number of dwelling units permitted by the underlying zoning. Receiving site density can exceed the base density of the underlying zone at the rate of one bonus unit for each unit transferred from a sending site. When TDR is used, maximum density may be 30 percent higher than base density or the development density recommended in the general plan, whichever is less. The approval of density bonus occurs through the PUD process, allowing the City to grant alternative design standards needed to achieve the bonus density as long as findings can be made that the project is compatible with surrounding development and promotes general plan policies.

Arlington County, Virginia, population 236,842 (2019), lies across the Potomac River from Washington DC. In 2006, Arlington County adopted a TDR ordinance allowing sending sites to be proposed throughout the county that provide open space, historic preservation, affordable housing, community recreation or public facilities. Unused sending site development capacity can be transferred to receiving sites using a special exception site plan process in which the county board determines whether or not the proposed transfer is consistent with land use plans, policies and ordinances. TDRs from residential sending sites can be converted to non-residential floor area at the rate of 3,000 square feet per foregone single family dwelling unit and 1,500 square feet per foregone multiple-family unit. The county subsequently adopted specific TDR zoning regulations with designated receiving sites for the Clarendon, Columbia Pike, and Fort Myers Heights districts aimed at affordable housing, historic preservation, green buildings, open space and other extraordinary benefits. To date, the county has approved at least four developments using TDR for historic preservation as well as the provision of open space and community/recreational facilities.

Arlington, Washington, population 19,483 (2019) is located 41 miles north of Seattle. In 2006, Arlington adopted Chapter 20.37 Transfer of Development Rights allowing for the designation of receiving sites within the city for the interjurisdictional transfer of TDRs from a 3,000-acre agricultural sending area within adjacent unincorporated Snohomish County. County Code 30.35A.050 specifies the number of TDR certificates issued when sending area land is preserved in compliance with County standards (See Snohomish County profile.)

The number of TDR certificates required for development approvals is as follows.

Type of Development	Number if Required TDR Certificates
Residential Subdivisions (single family)	25% of the number of lots included in the preliminary plat
All other residential development (multi-family)	50% of the number of residential units included in the official site plan
All nonresidential development	1 TDR Certificate per 10,000 square feet of gross floor area

Consequently, all development in the receiving area is subject to a TDR requirement even if the developer chooses to build at a low density. The first receiving area project did not develop as expected due to infrastructure constraints, but the TDR provisions remain in the Arlington Land Use Code as of 2020.

Aspen, Colorado, population 7,401 (2019), establishes floor area baselines within individual residential units. Owners of designated historic landmarks that have less floor area than baseline can sell one TDR per 250 square feet of unused floor area potential. Receiving site developers who choose the TDR option can exceed these baselines at the ratio of 250 square feet of floor area per TDR. Baselines vary between the seven zoning districts serving as receiving areas. On receiving sites zoned R-6, for example, single-family residential baselines range from 2,400 square feet of floor area on lots of 3,000 square feet or less up to 5,770 square feet of floor area on 50,000-square foot lots. By 2014, the program had created 64 TDRs and landed 24 TDRs at prices ranging from $175,000 to $240,000 per TDR.

Atlanta, Georgia, population 506,811 (2019), adopted TDR primarily to preserve historic landmarks, create greenspace and promote affordable housing. When the sending site is a designated landmark, the transferable development is the development potential of the site minus the landmark's existing floor area. The TDR program was hampered by high densities permitted as a matter of right by the zoning in Atlanta's Central Business District. However, baseline density is lower in Atlanta's Midtown area creating somewhat greater demand for transferable development rights. The program has preserved four historic landmarks including the house where author Margaret Mitchell once lived.

Austin, Texas, population 978,908 (2019), allows TDR in the form of bonus density for the off-site provision of affordable housing, historic preservation and open space. In certain cases, developers may pay a density transfer charge (DTC) in lieu of actually providing these community benefits. As of early 2020, Austin was considering a consultant's recommendation to increase the DTC in the CC zone under the affordable housing option from $10 to $12 per square foot of bonus floor area and to institute a DTC of between $12 and $18 per square foot for commercial developments which, as of February 2020, paid no DTC for affordable housing. The consultants based these recommendations on estimates of value increases created by bonus density and a goal of capturing 50 percent of that increment for public benefit. As of February 2020, 54 downtown projects were reported to have paid a total of $8.3 million to the Affordable Housing Trust Fund in DTC in lieu of building some or all of the affordable housing otherwise required in order to achieve bonus residential density.

Austin also offers a transfer mechanism aimed at providing flexibility in the siting of development while protecting the most sensitive portions of watersheds as critical water quality zones, often including flood plains and other areas closest to streams and lakes where impervious coverage limits can range from 30 percent to 65 percent. For example, in areas defined as Suburban Watersheds, 20,000 square feet of impervious cover can be transferred to an uplands zone for each acre of land in critical water quality zones that are dedicated to the city. In 2005, a city representative reported that these transfer provisions had been used many times but that the City had not attempted a formal inventory of the land preserved as a result.

Avon, Connecticut, population 18,302 (2019), lies ten miles west of Hartford and adopted a TDR ordinance in 2007 aimed at preserving natural resources, open space, and farmland. Sending areas are designated by map and allocated four TDRs per developable acre. Receiving areas can exceed a baseline of four units per acre to achieve a maximum density of eight units per acre by retiring one TDR per bonus unit. The 2016 Plan of Conservation and Development reported that the 2007 TDR mechanism had never been used and recommended consideration of program improvements including increasing the transfer ratio, reducing the minimum lot size of receiving sites, and expanding receiving areas.

Bainbridge Island, Washington, population 24,486 (2019), uses TDR to preserve wetlands, high vulnerability recharge zones, farmland, and open space. Adopted in 1996 and amended twice, the program has underperformed largely because the city allows density bonuses through alternative means that are cheaper and easier than TDR.

Bay County, Florida, population 174,705 (2019), surrounds Panama City on the Gulf of Mexico in Northwestern Florida's Panhandle Region. Chapter 32 of the Bay County Land Development Regulations aims to protect environmentally sensitive lands and minimize damage from floods, storms and hurricanes on Shell Island. Roughly half of the island is a state park, and the other half is an antiquated subdivision containing a mix of publicly- and privately-owned parcels.

In the sending area, lots of record can be allocated one TDR or the number of units allowed under zoning, whichever is greater. Receiving sites include parcels within 11 zones in the urban service area on the mainland. In 2006, one owner transferred rights on four Shell Island lots in return for the ability to gain additional floors within a proposed resort development.

Bay Harbor Islands, Florida population 5,793 (2019), is located on two islands with a total size of 240 acres created from mangrove swamps in Biscayne Bay, two miles north of Miami Beach. The Town's TDR program, as found in Section 23-22.2 of the town code, allows the owners of donor sites to transfer unused development potential to receiving sites in order to achieve desirable development projects as approved by the Town Council. As an example of a transaction, in 2007, a transfer was approved from a donor site capable of a maximum of 13 units but occupied by an eight-unit condominium built in 1959. The unused potential for five additional units was transferred to a receiving parcel, allowing construction of a 14-unit condominium building.

Beaufort County, South Carolina, population 192,122 (2019), adopted a TDR program in 2011 to reduce development potential within the Air Installation Compatible Use Zone (AICUZ), a quarter-mile buffer area surrounding Marine Corps Air Station - Beaufort. The number of TDRs available to a sending site is based on that parcel's development potential prior to the development limitations imposed by the AICUZ overlay. The receiving areas are lands under the jurisdiction of Beaufort County on Port Royal Island although the cities of Beaufort and Port Royal may also participate under the provisions of an inter-jurisdictional agreement. Any rezoning that increases residential density or commercial intensity potential also creates an overlay district in which developers must comply with TDR requirements if they choose to exceed baseline density or intensity. Baseline is the maximum density/intensity allowed under the prior zoning. When receiving area developers choose to exceed baseline, they must comply with TDR requirements by retiring one TDR for each three bonus dwelling units or every 5,000 square feet of non-residential floor area in excess of baseline. Developers also have the option of paying cash in lieu of each TDR that would otherwise be required according to an amount in the County's fee schedule. The proceeds of cash in lieu payments must be used exclusively to buy TDRs and pay for TDR program administration. If the County runs out of sending area landowners willing to sell their TDRs for full market value, the County can elect to create additional sending areas. Cash in lieu proceeds may be spent directly by the County or conveyed to a TDR bank for use under the provisions of an agreement with the bank.

Bellevue, Washington, population 148,164 (2019), is located between Lake Washington and Lake Sammamish, just east of downtown Seattle. In 2009, Bellevue adopted a TDR mechanism along with new zoning aimed at transforming its Bel-Red neighborhood from light-industrial uses to mixed-use neighborhoods that offer residents a walkable community with affordable housing, parks, restored streams, bicycling paths and transit. The TDR component is designed to inter-jurisdictionally transfer preserved land under King County jurisdiction that benefits Bel-Red and Bellevue as a whole including forests in the Mountains-to-Sound Greenway, farmland in the Snoqualmie Valley that serves the Bellevue's farmers markets, and the White River watershed, which supplies Bellevue's drinking water.

The transfer provisions are addressed in an interlocal agreement between Bellevue and King County in which the County pledges $750,000 to buy a parking lot and change a concrete-lined ditch to a natural stream within a park. In return, Bellevue agrees to accept 75 development rights representing the preservation of from 3,000 to 6,000 acres of land under County jurisdiction.

For each credit transferred from sending areas in the County, developers in Bel-Red will be able to build an additional 1,333 square feet of floor area or roughly one residential unit. Bonus floor area is obtained via a FAR Amenity System in five of Bel-Red's eight land use districts. In addition to TDR, the FAR Amenity System grants bonus density for affordable housing, park dedication, park improvements, trail dedications/easements, stream restoration and childcare/non-profit floor area. The FAR Amenity System allows density to go from base FAR 1.0 to maximum FAR 2.0 in two land use districts and from base FAR 1.0 to maximum FAR 4.0 in three other land use districts. Given the cap of 75 transferred credits, the program will allow an additional 99,975 square feet of bonus floor area in total.

Bellingham, Washington, population 92,314 (2019), uses interjurisdictional transfers to protect the watershed of Lake Whatcom which is located primarily in Whatcom County and supplies drinking water to much of the city and county. In at least three city neighborhoods, developers can exceed baseline density by making cash in lieu payments to the Lake Whatcom Watershed Property Acquisition Program. Developers who want to use TDR must make the in-lieu payment. In most other TDR programs offering cash-in-lieu options, the developer at least has a nominal choice of complying with cash or actual TDRs.

Belmont, California, population 26,941 (2019), uses both transfer of floor area and transfer of development rights provisions to encourage the permanent preservation of subdivided but undeveloped hillside land that is important to the city's character and also presents public safety hazards due to geologic problems and steep, potentially unstable slopes. On slopes greater than 45 degrees, the transfer of floor area mechanism allows as little as 900 square feet of total floor area including the garage, creating a substantial incentive for these owners to offer their lots as sending sites. The density transfer option allows one additional receiving site lot in exchange for the termination of three sending site lots, which is an unusual negative transfer ratio. However, these mechanisms allow property owners in antiquated subdivisions the opportunity for economic return from lots which would be difficult or impossible to develop under current subdivision standards. The owners of more than a dozen lots have taken advantage of these transfer options.

Berkeley Township, Ocean County, New Jersey, population 41,762 (2018) uses TDR to preserve wetlands and forested uplands in sending areas with vacant lots often given away with newspaper subscriptions over a century ago.

Bernards Township, Somerset County, New Jersey, population 27,205 (2018), maintains a farmland preservation TDR mechanism in its municipal code that allows transfers through a planned residential development process.

Berry Town, Dane County, Wisconsin, population 1,188 (2018) uses TDR to save farms and curb sprawl with receiving projects that promote planning goals.

Berthoud, Colorado, population 7,191 (2019), lies 45 miles north of Denver and 20 miles south of Fort Collins. In 1999, Berthoud tried to fashion a traditional TDR program but realized it would take time to resolve several issues. For example, there were disagreements about establishing desirable patterns of preservation versus growth, making it unlikely that the town would be able to quickly designate sending and receiving areas. Speed of adoption was important since Berthoud was anticipating large numbers of development applications.

Berthoud solved these problems with an ordinance that required developers to contribute to preservation in conjunction with zoning changes that increased density. Upzonings automatically created receiving areas in which the new zoning established baseline as the maximum density allowed under the prior zoning. Developers could exceed baseline by permanently preserving one acre of land or paying a density transfer charge (DTC) of $3,000 per additional single family residential unit or $1,500 per extra multiple-family residential dwelling unit. The town dedicated 94 percent of DTC revenues to the preservation of land and allowed the remaining six percent to cover administrative costs.

Using this approach, Berthoud eliminated the administrative complexity of traditional programs that require the issuance, transfer, and extinguishment of actual TDRs. Berthoud was able to target the highest priority properties for preservation since the town itself decided where to spend DTC revenue. Berthoud's DTC program also allowed the town the ability to quickly adopt a preservation mechanism without having to undergo the time-consuming and potentially contentious task of deciding what land deserves permanent protection and what areas are most appropriate for more intensive development. Economic analysis was necessary in order to develop the amount of land or DTC needed to produce a program that was economically viable for developers yet resulted in meaningful preservation. However, because the DTC option generated cash rather than TDRs, Berthoud was able to accomplish preservation regardless of fluctuations in sending site property values. Additionally, developers in Berthoud knew the maximum amount they would have to pay in advance, allowing certainty in the preparation of their budgets and relieving them of the time, uncertainty and effort involved in having to find and acquire actual TDRs. Berthoud suspended its program in 2009. As another benefit of Berthoud's approach, the town did not have to address the question of floating TDRs since the program did not use TDRs.

In 2002, the Berthoud approach was featured in an issue of the American Planning Association's Planning Advisory Service Memo using the nickname 'TDR-Less TDR'. This publication discusses the disadvantages as well as the advantages of DTC. As of 2021, at least 20 programs offered receiving site developers a DTC option and another six programs used DTC exclusively instead of actual TDRs. The nuts and bolts of building a DTC program are included in *The TDR Handbook: Designing and Implementing Transfer of Development Rights Programs*, published by Island Press.

Birmingham, Michigan, population 21,201 (2019), allows developments in three zones within its Triangle Overlay District to exceed baseline building height by conserving a designated historic landmark and transferring the unused floor area. Bonus floor area can also be achieved by four other means: LEED certification, mixed-use design, publicly accessible parking, and public plazas.

Birmingham Township, Chester County, Pennsylvania population 4,208 (2018), adopted a TDR program in 1978 to preserve farmland and environmental areas which has been hindered by a lack of demand to exceed baseline density.

Blacksburg, Virginia, population 44,303 (2019), allows density in its RR2 zone to increase from one to two units per acre by proffering of offsite open space.

Black Diamond, Washington, population 4,781 (2019), uses TDR to preserve public benefit lands with outstanding environmental, resource, recreational, community character, or public facility values. TDR allocation is doubled if the city accepts title to the sending site. The TDRs can be multiplied by three for sending sites designated as treasured places, meaning a public benefit site with such environmental, cultural, aesthetic, community, or strategic significance that immediate acquisition by the city is considered of utmost importance to the public welfare. The TDR bank sells TDRs for no less than appraised full market value.

Blaine County, Idaho, population 23,021 (2019), downzoned 19,290 acres of scenic agricultural land but allows property owners in this sending area to preserve their land and sell one TDR per 20 acres, the previous on-site density. Receiving areas developers can exceed baseline of one unit per 20 acres and achieve a maximum density of one unit per 2.5 acres by retiring one TDR per bonus unit. As of July 2021, 25 TDRs had been certified and 21 had been sold at prices ranging from $25,000 to $30,000 each. Using TDR, the Wood River Land Trust had permanently preserved its 131-acre Church Farm property.

Blue Earth County, Minnesota, population 67,653 (2019), lies 90 miles southwest of Minneapolis. The Agriculture and Conservation zoning districts, which apply to most county land, allow a baseline density of one residential unit per quarter-quarter section. However, property owners can transfer development rights from one quarter-quarter section to a contiguous quarter-quarter section if approved by conditional use permit (CUP). The sending and receiving sites do not have to be under common ownership, but approval of a CUP requires the Planning Commission to make findings regarding compatibility with surrounding property, consistency with county goals, and maintenance of environmental standards. Since these transfers occur at a one-to-one ratio, the overall development level does not change and, ideally, the transfers result in preservation of the most productive agricultural land and the most sensitive environmental areas. At least 5,000 acres have been protected by this program.

Bluffton, South Carolina, population 20,799 (2019), uses TDR to protect open space, environmental areas, and important waters. As of 2014, the program experienced at least eight transfers, mostly to the Town's TDR bank. In 2012, the owners of Palmetto Bluff agreed to transfer 1,300 units from land near the headwaters of May River, a defining feature of the Town and an important shellfish habitat.

Boulder County, Colorado, population 326,196 (2019), begins roughly 15 miles northwest of Denver's downtown. The eastern third of the county extends into the Great Plains and the western two thirds encompasses portions of the Rocky Mountains with extensive public lands including the Roosevelt National Forest, Rocky Mountain National Park, and several other parks, preserves, and wilderness areas.

Boulder County is an open space super star. As profiled in Lasting Value: Open Space Planning and Preservation Successes, in addition to 137,308 acres held by the federal government, Boulder County itself protects 105,386 acres by ownership, lease and easement. The county uses various preservation techniques including voter-approved open space sales taxes, property taxes, grants, and TDR mechanisms that feature a remarkable amount of voluntary intergovernmental cooperation.

In 1981, Boulder County began motivating open space preservation by adopting a clustering option known as non-urban planned unit development, or NUPUD. This mechanism allows base density to be doubled, from one unit per 35 acres to two units per 35 acres, when development is concentrated on no more than 25 percent of the parcel and the remaining 75 percent is preserved by a permanent conservation easement.

In 1989, Boulder County added a modification that created a transfer of development rights option known as non-contiguous non-urban planned unit development, or NCNUPUD. In this process, one PUD application allows the transfer of development potential between two non-contiguous parcels.

According to Article 6 of the Boulder County Land Use Code as of June 2019, Planned Unit Developments, the sending and receiving parcels of a NCNUPUD must be located within the same water service area and/or water and/or sanitary district, school district, and fire district response area unless mitigation measures are adopted to address development imbalances. On sending sites, most of the land proposed for preservation must be identified in Boulder County's Comprehensive Plan as having agricultural, natural, cultural, environmental, scenic, community buffer or open space significance. The receiving sites cannot be located within areas likely to be annexed by a municipality without the approval of that jurisdiction.

NCNUPUD can triple the density permitted by NUPUD. For example, a 35-acre parcel is allowed one unit as a matter of right, two units when the NUPUD option is used, or six units via NCNUPUD. No more than half of the receiving site can be developed. If the receiving site contains agricultural lands of national importance, no more than 25 percent of the parcel may be developed. If at least half of the units are transferred in the original transfer, the remaining units can be held for use in subsequent applications as long as various conditions are met.

Boulder County added a second TDR tool called Transferred Development Rights Planned Unit Development (TDR/PUD) with the purpose of preserving generally contiguous properties to protect agriculture, rural character, open space, scenic vistas, natural features, and environmental resources. The county maps some sending areas as well as the sending and receiving areas for Niwot, an unincorporated urbanized area. In addition, sending areas can be designated by intergovernmental agreements between the county and municipalities. The code allows the transfer of various types of development potential: residential and yard area minimums. The code also permits conversions from residential to non-residential uses. At least 75 percent of the development transferred to a receiving site must come from a sending site within the same designated subarea unless the county board grants an exception. Receiving sites must be located adjacent to an arterial, collector, or transit route and the receiving site development must be compatible with adjacent development.

Sending sites must be protected by a conservation easement on the entire parcel and no more than five percent of the sending property may be developed. Subject to a public hearing, TDRs can be severed and sold from land acquired in fee by a governmental entity or protected by an easement held by a governmental entity. The amount of development allowed to remain on sending sites versus transferred to a receiving site varies based on the size of the sending property. For parcels between 35 and 52.49 acres, two development rights can be transferred, or one unit can be transferred, and the other unit built on site as long as on-site construction is consistent with the comprehensive plan. In the last of seven categories, parcels 140 acres and larger can transfer two TDRs per 35 acres or any combination which does not exceed a transfer of two units per 35 acres or one unit per 70 acres on site.

Boulder County allows property owners with deliverable agricultural water rights to transfer one additional unit per 35 acres if the owner provides an undivided interest in those water rights to the county.

Conservation easements on sending sites state that the easement can be terminated if termination is consistent with the comprehensive and management plans of Boulder County and other interested governmental entities. However, the county may require compensation for easement termination in an amount sufficient to offset any loss of public benefit.

As explained above, the TDR/PUD process can work inter-jurisdictionally as well as intra-jurisdictionally. Boulder County uses intergovernmental agreements (IGA) to facilitate the transfer of development potential from sending sites within unincorporated Boulder County to receiving sites within incorporated cities. This process was tested in the Boulder Valley TDR Program, implemented through an intergovernmental agreement between the City of Boulder and Boulder County adopted in 1995. At last count, Boulder County had signed IGAs with eight municipalities and the unincorporated town of Niwot.

Typically, a planning area is created around the community participating in the TDR program with Boulder County. In this way, the community can maximize the benefits of open space preservation by requiring that a minimum percent of transferred development rights come from the rural areas immediately surrounding them. The planning area may be further divided into subareas with the transfers required to occur between sending and receiving sites within that subarea. With this technique, the neighborhoods accepting the extra receiving site development also are closest to the open space which that extra density made possible.

While certain procedures remain constant, individual requirements are spelled out by the separate IGAs. The most detailed IGA, adopted in April 1995 as the "Boulder Valley TDR Comprehensive Development Plan", is between Boulder County and the City of Boulder.

In general, this IGA combines the City of Boulder's commitment to accept transferred development rights from the County and the County's commitment to preserve rural character. More specifically, the County agrees not to approve NCNUPUDs within the Plan Area or NUPUDs within the Planning Reserve Area unless they are jointly approved by both the City and County. Correspondingly, the City agrees not to annex unincorporated land or otherwise allow development that is contrary to the jointly-adopted Boulder Valley Comprehensive Plan (BVCP).

The Boulder Valley Comprehensive Plan (BVCP) identifies four categories of sending sites: 1) the Rural Preservation Area; 2) the Accelerated Open Space Acquisition Area; 3) the Northern Tier Lands; and 4) Private Land Enclaves lying between the Boulder Mountain Parks and the Arapaho-Roosevelt National Forest west of the City. Receiving sites can include land within the boundaries of the City of Boulder's community service area, areas being annexed to the City in accordance with the provisions of the BVCP or lands within a rural planning area that have been approved under the provisions of the BVCP. Another area, the Planning Reserve Area, is suitable for urban development in the long-term future; but in the near term it is planned to remain rural and cannot be used as a receiving site.

Under the IGA, the City of Boulder agrees to accept up to 250 development rights. Certificates of

Development Rights are issued only after a conservation easement, precluding further development and granted jointly to the County and City, is recorded for the sending site. The IGA further requires the City and County to establish a joint committee to monitor the progress of the TDR program. The Agreement also prevents the County from making any changes to its NUPUD or NCNUPUD regulations without the City's consent. Finally, the IGA between Boulder City and Boulder County terminates five years after its effective date.

The IGA between the County and the City of Lafayette, which became effective in December of 1995, is similar to the IGA with the City of Boulder. However, this IGA does not limit the number of development rights which can be transferred to Lafayette. In addition, this TDR plan clearly limits the sending sites to designated areas within a 27-square-mile Plan Area that extends from one to four miles in each direction from the Lafayette City Limits. The IGA with the City of Longmont, which became effective in January of 1996, similarly limits the sending area to about 50 square miles at the periphery of the Longmont TDR Planning Area.

As of 2009, TDR in Boulder County had preserved between 4,400 and 5,900 acres (Pruetz & Standridge 2009). However, TDR activity had declined by 2005 because there were fewer and fewer parcels of 35 acres or greater that qualified as sending sites. At that time, over 85 percent of county land was either within incorporated communities or already preserved by public ownership or conservation easements resulting from federal and state programs as well as the many preservation options launched by Boulder County and its municipalities (Fogg 2005).

TDC Program

In 2008, Boulder County adopted a new program which requires developers to acquire transferable development credits (TDCs) to exceed a floor area baseline of 6,000 square feet within individual dwelling units. This 6,000-square foot baseline includes garages, basements and residential accessory structures. Each TDC allows an additional 500 square feet of floor area until reaching 1,500 bonus square feet of floor area; above that 1,500 square foot increment, two TDCs are needed for every 500 square feet of additional floor area. (Under limited circumstances, TDRs from the TDR/PUD program discussed above can be converted to TDCs.)

Sending sites for the TDC program are legal building lots with legal access. Property owners who choose to participate receive the following TDCs depending on the permanent restrictions they record on their properties:

- 2000 square feet of floor area or less: 2 TDCs
- 1500 square feet or less: 3 TDCs
- 1000 square feet or less: 4 TDCs
- Keep land vacant: 5 TDCs for mountains properties and 10 TDCs for plains properties.

This program aims to preserve rural character, motivate affordable housing, and promote more sustainable development by offsetting the impacts of larger houses with the preservation of smaller houses and the open space resulting from smaller homes. As of January 2021, 251 TDCs had been sold at prices ranging from $3,000 to $10,000 per TDC, with most sales occurring toward the high end of that range.

References

Fogg, P. 2005. Correspondence with author: February 9, 2005.

Pruetz, R. and N. Standridge. 2009. What Makes Transfer of Development Rights Work? Success Factors From Research and Practice. Journal of the American Planning Association. Vol. 75, No. 1, Winter 2009.

Brentwood, California, population 61,961 (2019) uses mitigation requirements, in lieu fees, and transfer of agricultural credits to preserve farmland. At least one transfer has occurred. The Agricultural Preservation Program, Chapter 17.730, specifies that if a court of law finds that the public interests of an agricultural conservation easement can no longer be fulfilled, the interest in the agricultural mitigation land may be extinguished through sale, with the proceeds redeposited into the city's mitigation fund and used for continued agricultural preservation.

Brevard County, Florida, population 601,942 (2019) uses TDR to preserve agricultural land, oceanfront land, wetlands and other environmentally significant areas. Under one TDR option, developers can use the planned unit development mechanism to relocate growth while maintaining the combined development potential of the sending and receiving sites. Under another option, TDRs can be used to shift development potential between designated areas, allowing density on receiving sites to grow by up to 20 percent.

Brisbane, California, population 4,671 (2019), uses TDR to protect environmentally sensitive, hillside land that includes habitat for two federally endangered butterfly species. For each 20,000 square feet of permanently preserved sending area land, receiving sites can exceed a baseline density of one unit per 20,000 square feet and create a bonus lot of at least 5,000 square feet. In January 2005, the Brisbane City Council adopted its first TDR application, allowing two additional single-family dwellings in exchange for City acquisition of an acre of open space near the ridgeline of San Bruno Mountain.

Buckingham Township, Bucks County, Pennsylvania, population 20,260 (2018), became one of the first communities to use TDR for farmland preservation in 1975. Land in the preservation district was downzoned from one unit per acre to either 0.2, 0.3 or 0.5 units per acre for on-site development. As compensation for this downzoning, sending site owners were allowed to transfer development rights at the rate of one unit per acre. If sending site owners opted to build on sending sites, they were required to cluster new development on either 10 or 20 percent of the sending site and deed-restrict the remainder for agriculture. Land in the development district was up zoned to a 2.5 units per acre baseline that developers could exceed by TDR. Initially, the program was hindered by weak demand to exceed baseline density and the need for receiving area infrastructure. The township subsequently addressed these issues and, as of 2020, had saved a total of 5,312 acres of farmland of which 505 acres were preserved by TDR.

Burbank, California, population 103,703 (2019), uses TDR to allow flexibility in the concentration of development while keeping the overall density/intensity of a specific plan area within transportation system capacity. Properties transfer housing units or non-residential floor area on a one-to-one basis using trip generation rates to convert between different land uses. A conditional use permit can approve transfers that promote compatibility and planning goals. The program resulted in appropriately scaled buildings at key locations as envisioned.

Calvert County, Maryland, population 92,525 (2019), lies 35 miles southeast of Washington, D.C., on a peninsula between the Patuxent River and Chesapeake Bay. The TDR program here was initially adopted in 1978, making it the state's first TDR program. It was primarily designed to preserve farms and forests, but staff recently noted that Calvert County uses TDR and land preservation in general as a climate action tool that mitigates the GHG emissions caused by sprawl, protects local food sources, and helps property owners adapt to wildfire, flood, storm surge, sea level rise and other hazards exacerbated by climate change.

Three zoning districts that account for most of the land area in Calvert County can become either sending or receiving sites. Owners who wish to become sending sites apply for a designation of Agricultural Preservation District (APD). If approved as APD, the county certifies the number of TDRs allocated to the property. In most instances, the allocation is one TDR per acre minus five TDRs for each existing dwelling. The APD status is in effect for at least five years. During that time, the site must be actively farmed or forested. If no TDRs are sold after the five-year period, the owner may remove the APD designation. However, after the first TDR is sold, a permanent easement must be placed on the entire property. The number of residential lots allowed by the permanent easement varies depending on the size of the APD:

- Less than 25 acres, no additional lots in addition to the existing house;
- At least 25 but less than 50 acres, one lot in addition to the existing house;
- At least 50 but less than 75 acres, two lots in addition to the existing house;
- 75 or more acres, three lots in addition to the existing house.

Alternatively, owners of land in these three zoning districts can apply to become receiving sites and increase allowable density at the ratio of five TDRs per bonus residential lot. In the Farm & Forest and Rural Community districts, baseline density is now one residential lot per 20 acres (with exceptions), and the use of TDR can double that density to one lot per ten acres. In the Residential District, baseline density is one unit per four acres (with exceptions) and the density allowed by TDR depends on the receiving site's proximity to a town center. Receiving sites in the Residential District that are outside a one-mile radius of any Town Center can double baseline density to one lot per two acres with the retirement of five TDRs for each additional lot. Receiving sites in the Residential District within a one-mile radius of a Town Center can achieve a maximum density of four lots or units per acre, which represents a 16-fold increase.

In the county's Town Center districts, land can only serve as receiving sites. In these districts, five TDRs are required for each dwelling unit in excess of one unit per acre. The baselines and maximum with-TDR densities and lot sizes vary depending on which of the seven Town Centers the receiving area is located in and the nature of the project, meaning single family, duplex, townhouse or multi-family residential. To offer one of the simpler examples, in the Dunkirk Town Center, baseline is one dwelling unit per acre, but TDRs can be used to increase single-family, duplex, and townhouse developments to four units per acre and TDRs can boost multiple-family residential developments to 14 units per acre.

Calvert County has adopted major revisions to its program since 1978. In 1993, the county downzoned land in the Farm & Forest, Rural Community, Residential and Town Center districts. After another downzoning in 2003, baseline density in the Farm & Forest and Rural Community districts had dropped from one unit per five acres in 1978 to the current baseline of one unit per 20 acres. These downzonings increased the motivation of the owners of productive resource lands to become sending sites rather than develop to the relatively low baseline density allowed on site. The downzonings likewise encouraged owners of property that qualified to become receiving sites to use the TDR option in order to exceed the lower baselines allowed after these downzonings. Studies of this program attribute much of the success of the Calvert County program to these downzonings (McConnell, Kopits & Walls 2003; Walls 2012).

Calvert County is an active participant in the TDR market. In 1993, the county started the Purchase and Retirement (PAR) program, which buys and retires TDRs. In a second program adopted in 2001, Leverage and Retire (LAR), farmers preserve their land and are reimbursed by the County over time. For example, a landowner might receive tax-free interest payments over 15 years followed by a payment for principal at the end of 15 years. The County benefits by being able to protect more land with limited near-term expenditure. However, landowners also benefit by deferring income into years when they plan to be retired and earning less income from other sources. The PAR and LAR programs contributed to price stability over time. But observers also note that the purchase price paid by the county can also inhibit private market sales when sending site owners decline to sell to developers because they are hoping to be able to sell their TDRs to the county for a higher price. In 2018, when the real estate market fully recovered from the Great Recession, 238.5 TDRs sold at an average price of $6,043.

A viable private market is needed in order to absorb the large supply of TDRs in Calvert County. As of August 2020, 12,000 TDRs were reported to be available for sale. Of this total, 8,000 could no longer be reattached to sending sites because at least one TDR has been sold and a permanent easement recorded. Because no TDRs had yet been sold from other sending sites, the remaining 4,000 TDRs could be removed from the program and become available again for on-site use after completion of the required time period. As of 2020, county staff were exploring potential code changes aimed at increasing the demand for TDRs (Myers 2020).

Outcome

By 2019, the TDR program in Calvert County had preserved over 23,000 acres, which is over half of the county's 40,000-acre goal. Because land in three zoning districts can become receiving sites as well as sending sites, the program has been criticized for contributing to the fragmentation of agricultural areas. However, the density allowed via TDR in the two largest zones is one unit per ten acres, a density that program supporters see as a reasonable tradeoff for accomplishing the amount of permanent preservation achieved in Calvert County.

References
McConnell, Virginia, Elizabeth Kopits, and Margaret Walls. 2003. How Well Can Markets for Development Rights Work? Evaluating a Farmland Preservation Program. Washington, D.C.: Resources for the Future.

Myers, Dick. 2020. Planners Deny TDR Changes. Calvert County Times Newspaper. August 31, 2020.

Walls, Margaret. 2012. Markets for Development Rights: Lessons Learned from Three Decades of a TDR Program. Washington, D.C.: Resources for the Future.

Cambridge, Massachusetts, population 116,632 (2019), uses TDR in three neighborhood overlay districts to create parks and housing at sending sites and increase density at receiving sites near transit.

Cape Elizabeth, Maine, population 9,313 (2018), has a TDR ordinance aimed at preserving, farmland, greenbelts, potential recreational sites, wildlife habitat and other places of scenic or cultural significance.

Caroline County, Maryland, population 33,049 (2019), adopted TDR in 1977 to preserve agricultural and conservation sending areas. Participating sending area owners must demonstrate that their parcels can accommodate development using percolation and hydrologic tests. One development right must be retained on sending parcels but owners can sell some or all of the site's other TDRs. As of 2016, this program had preserved 2,827 acres.

Carroll County, Maryland, population 168,447 (2019), uses TDR to protect land with viable mineral deposits, particularly a form of marble used to make building materials. If the owners of potential sending sites cannot use clustering to avoid a Viable Resource Overlay, they can transfer unused development potential at the ratio of two TDRs for every dwelling unit otherwise permitted by the underlying zoning. Receiving sites are areas zoned R-40,000, R-20,000 and R-10,000, where cluster subdivisions can increase density at two TDRs for every ten lots. The TDR provisions have been used at least once, permanently preserving an 80-acre parcel.

Carver, Massachusetts, population 11,777 (2018), uses TDR to preserve farmland, open space, historic landmarks and environmental resources. Receiving sites have a baseline of six units per net useable acre. TDRs are multiplied by 1.5 for affordable housing and by 2.0 for age-restricted affordable housing.

Cave Creek, Arizona, population 5,760 (2018), uses TDR to permanently preserve consolidated open space by allowing receiving sites to satisfy open space requirements on sending sites in compliance with the general plan.

Cecil County, Maryland, population 102,855 (2019), uses TDR to preserve farmland and environmental areas. Four receiving zones offer special development regulations as well as bonus density to projects using TDR.

Central Pine Barrens, New York - The Central Pine Barrens lies in Suffolk County, New York, which includes the entire eastern end of Long Island and had a population of 1,477,000 in 2019. The Pine Barrens is the largest single undeveloped area on Long Island and is home to pitch pine and pine-oak forests, coastal plain ponds, marshes and streams which provide open space and various outdoor recreational opportunities. The area protects the largest concentration of endangered, threatened and special concern plant and animal species in the State of New York, including dwarf pines. In addition, the Pine Barrens constitute the deep recharge area for one of the largest sources of groundwater in New York State, an aquifer that provides drinking water for 1.6 million people. This unique area is protected by a regional plan implemented by various preservation strategies including TDR. The entire profile of this TDR program can be found in Chapter 9 of Part I.

Chanceford Township, York County, Pennsylvania, population 6,206 (2018), has had a farmland TDR mechanism in its zoning code since the late 1980s.

Chapel Hill, North Carolina, population 60,998 (2019), uses TDR to permanently preserve watersheds for water quality, floodwater management, and habitat protection by allowing density transfers to receiving sites in twelve zones.

Charles County, Maryland, population 163,257 (2019), is located 25 miles south of Washington, D.C., across the Potomac River from Virginia. Agricultural preservation has been a goal of Charles County for decades. In 1980, the County created an Agricultural Land Preservation Program sponsored by the Maryland Agricultural Land Preservation Foundation (MALPF). When farmers enroll in this program, they cannot develop their farms for five years. In return, the farmers receive tax credits and protection from nuisance complaints registered against normal farming practices by non-farm neighbors. The program also offers to purchase development rights in order to permanently preserve land for agriculture.

Despite these agricultural preservation efforts, the County continued to lose farmland to development. In the 1980s, the Agricultural Preservation Program purchased development rights on 223 acres of County farmland. However, in the five-year period from 1982 to 1987, Charles County farmland declined from 84,000 acres to 68,000 acres, a loss of 16,000 acres. Consequently, when the Agricultural Preservation Element of the County's Comprehensive Plan was adopted, it recommended the establishment of a transferable development rights program as an additional means of promoting farmland preservation.

In 1991, a consultant's study concluded that TDR could be effective in preserving farmland. However, rather than adopting a TDR ordinance for all of the agricultural land in the County, this study recommended a program that would apply to a more manageable amount of farmland. Charles County followed up on that recommendation by adopting an ordinance in 1992 which allows for the transfer of development rights only from land enrolled in MALPF Agricultural Land Preservation Districts. The description in the process section below is based on Article XVII Transferable Development Rights (TDRs) in Designated Agricultural Land Preservation Districts as that article appeared in the Charles County Zoning Regulations in November 2011. As of January 2021, ecode360 indicates that this article has not been amended since 1996.

Process

In the Charles County program, sending areas are properties established as County Agricultural Preservation Districts under the program sponsored by MALPF. To be considered for a Preservation District, at least half of the land area of a farm must contain Class 1, 2 and 3 soils and the farm must have soil/water conservation improvements in place. As mentioned above, farmers who enroll in the Agricultural Preservation District Program cannot develop the property for a five-year period. After five years, the farmer can choose to withdraw.

The zoning code allows one dwelling unit per three acres. However, as a practical matter, it is difficult to actually obtain this density due to regulations regarding forest management, storm water control, groundwater protection and resource preservation. As a result, the achievable density is often one unit per five acres.

However, as an incentive to transfer development rights rather than develop on sending sites, transferable development rights are allocated at the rate of one TDR per three acres of land in an Agricultural Preservation District. Since the density allowed under the zoning code is one dwelling unit per three acres, the program offers only a one-to-one transfer ratio in theory. But, due to the difficulties mentioned above regarding building on sending sites, the program offers a practical transfer ratio that is actually slightly higher than one-to-one.

If the sending site owner chooses to proceed, a covenant is prepared and recorded specifying the number of development rights to be transferred. The sale of any of these rights immediately encumbers all TDRs assigned to the sending site. These covenants are contained in a document known as an instrument of original transfer. This instrument also includes covenants documenting that the development rights have been transferred from the transferor to the transferee.

Severed development rights do not have to be transferred directly to a receiving site. The sending site owner can transfer these rights at any time and to any person the way any other property interests can be transferred. For assessment and taxation purposes, severed TDRs are deemed appurtenant to the sending parcel until they are approved for use at a specific receiving site.

Many programs allow TDRs to be reattached to sending site if the site has been determined no longer suitable for conservation by a new comprehensive plan and/or a court decision. In contrast, Charles County allows sending site owners to buy back their development rights with few conditions other than that the sending site cannot ultimately have more than its original number of development rights.

The potential receiving sites are parcels zoned RL, RM, RH, MX, TOD, PRD, CER, CRR, CMR, AUC and WC. In the RL, RM, RH, MX, TOD and PRD zones. TDRs can only be allowed by using cluster or planned development standards. The owners of land in these zones can use transferred development rights to exceed the density allowed in each zone as a matter of right up to the maximum density limits for each zone. Additional density bonuses are also available when at least 10 percent of the units in a receiving site development meet affordable housing standards. Consequently, Figure V-1 of the Zoning Regulations establishes a progression of bonuses. For example, the RL (Low Density Residential) zone has the following progression: baseline density - one unit per acre; with affordable housing density bonus – 1.22 units per acre; with maximum TDRs – 3 units per acre; and with maximum TDRs and the affordable housing density bonus – 3.22 units per acre.

Program Status

In a 2007 study for Resources for the Future, Margaret Walls and Virginia McConnell concluded that the Charles County TDR program has both supply and demand problems (Walls & McConnell 2007). On the supply side, sending area land must be enrolled in the MALPF program, a requirement that not all farms can meet. The MALPF program pays about 75 percent of market value for a permanent easement, a level of compensation which is generally higher than the price that farmers would get by selling their TDRs. Consequently, landowners prefer to wait for MALPF funding even though these funds are limited.

Demand for TDRs has also been low because of preferences for low density development in Charles County. Walls and McConnell found that most TDR purchases have occurred in the lower density receiving areas and that developers who used TDR rarely used the full density potential available to them. Consequently, even at low TDR prices, developers in Charles County appear to have little incentive to use this tool.

A 2009 study reported that TDRs had preserved over 4,000 acres (Pruetz & Standridge 2009). In 2010, Charles County added Activity Urban Center (AUC) and Waldorf Center (WC) zoning districts that allow high density residential development supported by public transportation. Waldorf, an urbanizing area in northern Charles County is only 28 miles south of Washington, DC. The AUC and WC zones have a residential baseline of 12 units to the acre and allow developers to use TDR to achieve higher densities. The willingness to continue using TDR in new zoning districts suggests that Charles County still finds TDR to be a useful preservation tool. As of 2016, the TDR program in Charles County had preserved 5,274 acres (Maryland 2016).

References

Maryland. 2016. Transfer of Development Rights Committee Report. Maryland Department of Planning. Accessed 1-29-21 at TDR-committee-report-2016.pdf (maryland.gov).

Pruetz, R. & N. Standridge. 2009. What Makes Transfer of Development Rights Work? Journal of the American Planning Association. Vol. 75. no.1.

Walls, M. & V. McConnell. 2007. Transfer of Development Rights in U.S. Communities: Evaluating Program Design, Implementation, and Outcomes. Washington, D.C.: Resources for the Future.

Charlotte County, Florida, population 188,910 (2019), lies on Florida's Gulf Coast between Tampa and Fort Myers. The West region of the County includes miles of mangrove shoreline along the edges of the Myakka River and Charlotte Harbor as well as the barrier islands of Manasota Key and Gasparilla Island. The interior of Charlotte County is primarily composed of small lakes, swampland, and agricultural land. Punta Gorda, population 19,571 (2019), is the only incorporated city but many developed subdivisions are located throughout the unincorporated portion of the county. The majority of the land in the West and Mid regions of the County along with some coastal areas of the South/East region were subdivided beginning in the 1880's. Much of these platted lands remain vacant and the County is forced to retro-plan for services, infrastructure, and growth management. Due to the prolific subdivision of land, the County contains tens of thousands of units of potential residential density in the form of existing vacant lots with single-family and multi-family zoning designations.

In 1988, the County adopted a Comprehensive Plan that called for the preservation of natural, historical, archeological, and cultural resources. To accomplish that goal, the County created land use restrictions that the County acknowledged could be burdensome to landowners. To help alleviate this burden, the County adopted an ordinance in 1994 which allowed development rights to be severed from restricted properties and transferred to more suitable areas. TDR ordinance amendments adopted in 2001 allowed receiving zone developers to make a payment-in-lieu-of-transfer into a County Land Acquisition Trust Fund. Another amendment in 2004, changed the name of the program to Transfers of Density Units (TDU). The TDU provisions were amended again in 2018 and as of 2021 appear in Section 3-9-150 of the zoning code.

The 2018 amendment reiterated that the program aims to redirect development away from sending areas with environmentally sensitive, historic, agricultural, and archeological resources. Properties qualify for the sending zone (SZ) designation by having at least one of 12 criteria, including coastal high hazard areas (CHHA), prime aquifer recharge areas, public water system protection areas, critical wildlife corridors, and managed neighborhoods, the county's term for undeveloped or underdeveloped platted lands, often containing or adjacent to sensitive environmental resources, that the county comprehensive plan has designated by map and where further development is discouraged;

The code has several rules governing how much density must be or can be eliminated from sending sites. Within many resource categories, the county allows or requires retention of one unit of density under various conditions. For example, development potential must be completely eliminated from sending parcels within managed neighborhoods, however, one unit of density must be retained for parcels designated as managed neighborhood if they have public water or sewer service.

Generally, preservation is accomplished by covenant. However, owners may propose that the county purchase or assume ownership of a sending site by donation. If the county agrees to this proposal, the sending site owner does not have to execute a covenant. Otherwise, the sending site property owner must submit a covenant that protects and maintains resources important to the county. Covenants must include a management plan when sending sites contain environmentally sensitive, historic, archeological, agricultural, or wildlife corridor resources. Management plans are not required for substandard parcels in the Coastal High Hazard Area (CHHA). In the SZ, the transferable density is the base density.

To qualify as a receiving zone (RZ), properties must be located within the urban service area and within any of seven mapped land use classifications: emerging neighborhood, maturing neighborhood, economic center, economic corridor, redevelopment area, revitalizing neighborhood, and rural settlement overlay. Outside the urban service area, the RZ can also be used in places designated as rural community mixed use. However, approval of an RZ is expressly prohibited on land with seven characteristics including managed neighborhoods, resource conservation/preservation areas, barrier islands (unless the transfer is between properties of the same island), and properties with the resources that would likely qualify the property as a sending site.

The code is particularly specific about transfers involving land in the CHHA. Generally, receiving sites in the CHHA can only receive density from sending sites in the CHHA that are more vulnerable to flooding than the receiving site. However, exceptions to this general rule apply in some areas.

Charlotte County allows developers the option of transferring actual density units or contributing to the county's Land Acquisition Trust Fund (LATF). The per-unit price under the LATF option is determined at the time the county board approves the TDU application. Upon approval of the TDU, the applicant can choose to pay the LATF per-unit price established when the TDU application was approved. Alternatively, the applicant can pay later according to a process that requires payment in an amount that reflects the cost of land acquisition at the time payment is made. Payments to the LATF occur prior to preliminary plat approval or, if plat approval is not needed, prior to building permit issuance. The LATF option is not available for applications to transfer density onto receiving sites in a tropical storm surge or category 1 hurricane storm surge zone as well as receiving zones in two other locations.

A study of 1,577 density bonus units granted between 2003 and 2005, suggests that the developers of larger projects prefer to acquire actual TDUs while the builders of smaller projects tend to use the LATF option. Specifically, only three of the nine developments in this study used LATF, paying $3,700 each in lieu of acquiring four actual TDUs (Williams 2005).

As of 2014, Charlotte County had approved 52 TDU applications, severed 14,368 TDUs and protected 1,818 acres. Considering the high number of TDUs, the protected acreage may seem relatively small. However, many of these TDUs represent the protection of tiny lots in antiquated subdivisions created by land schemes in the 1950s and 1960s (Linkous & Chapin 2014).

References
Linkous, E. & T. Chapin. 2014. TDR Performance in Florida. Journal of the American Planning Association. Volume 80. 2014. Issue 3.
Williams, I. 2005. Correspondence with author.

Chattahoochee Hills, Georgia, population 2,867 (2018), uses TDR to conserve natural, agricultural, environmental, historical, cultural and open space resources. Developers of receiving site projects can gain bonus density with actual TDRs or by paying a Density Transfer Charge that the city uses to preserve sending sites.

Chesterfield Township, Burlington County, New Jersey, population 7,500 (2018), is located within the I-95 megalopolis, 30 miles northeast from Philadelphia and 70 miles southwest from New York City. Despite being in the path of intensive growth pressure, Chesterfield has combined state and county preservation funds with its own TDR program to preserve 7,956 acres of farmland to date, which represents most of its agricultural land and over half of its total land area.

In 1975, Chesterfield Township adopted a TDR ordinance that went unused and was amended in 1985 and 1987. After the State of New Jersey adopted the Burlington County TDR Demonstration Act in 1989, Chesterfield developed several draft plans, ultimately adopting a new TDR ordinance that was amended in 2002 along with a master plan for Old York Village, the 560-acre TDR receiving area.

In contrast with incremental TDR programs that evolve over time, Chesterfield meticulously developed its program to maximize effectiveness. TDR supply was calculated by analyzing soil suitability for all qualified sending sites. Old York, the program's only receiving area, was planned as a traditional neighborhood design (TND) development with a density comparable to the nearby historic village of Crosswicks. To reduce dependency on car travel, Old York has local-serving retail, offices, recreational sites, public facilities and an elementary school plus walking paths connecting the various neighborhoods with each other as well as Crosswicks. This attention to detail required extraordinary public involvement and active participation from the development community. By doing this legwork in advance, Chesterfield was able to adopt detailed development requirements and architectural design standards that eliminated the need for public hearings as long as builders followed the rules and transferred the required number of TDRs.

The sending area consists of the 7,472 acres of farmland in parcels over 10 acres in size. The sending area is zoned AG, which nominally allows on site development at a density of one unit per ten acres or one dwelling per 3.3 acres when owners cluster development and preserve at least half of the site for agriculture. The TDR program allocates TDRs to sending area land based on the suitability of each property for development on septic systems: slight limitation 1 TDR/2.7 acres; moderate limitation 1 TDR per 6 acres; severe limitation 1 TDR per 50 acres. After adding a ten percent bonus allocation to increase motivation to use the TDR program, the Township calculated a total supply of 1,383.25 TDRs in the sending area. When averaged over the 7,472 acres of sending area land, the allocation adds up to roughly one TDR per five acres.

When owners choose to voluntarily record easements on this land, future development is limited to a maximum density of one dwelling unit per 50 acres. The easements are permanent. However, Chesterfield allows TDRs to be reattached to previously preserved sending sites if the holders of TDRs are unable to sell their TDRs and the planning board concludes that the TDRs are not likely to be purchased in the near-term future.

The receiving area of Old York Village is zoned Planned Village Development and was designed with the expressed purpose of accommodating all of the TDRs from the 7,472 acres of farmland in the sending area. At build-out, Old York Village is planned to accommodate 1,200 residential units, of which 72 percent will be on lots ranging from 7,000 to 10,000 square feet in size. Another 20 percent will be triplexes and two percent apartments with 40,000 square feet of commercial floor area and an elementary school.

In an unusual arrangement, TDR requirements apply to all private development in Old York with the exception of affordable housing units. The allowance rates are 1 TDR per detached perimeter village lot, 0.9 TDRs per detached village lot, 0.75 TDRs for a triplex dwelling lot, 0.5 TDRs for a condominium/apartment unit, no TDRs for low- and moderate income housing units (six percent of all units must be low-moderate income units), 1 TDR per 2,000 square feet of retail/office, 0.5 TDRs per home office, 1 TDR per acre of private institutional use (cemetery, private outdoor recreation), 1 TDR per 2,000 square feet of non-public institutional floor area (houses of worship and child care centers), and no TDRs for public buildings including public schools, libraries, and municipal facilities.

Chesterfield uses the Burlington County Development Credit Bank to facilitate marketing and transfers. The Township and Burlington County each provide half of the capitalization for the bank's activities in Chesterfield. The bank buys TDRs at prices established by the bank board taking into consideration the most recent private sales of TDR in arms-length transactions not involving the bank. To avoid competing with private sellers, the bank can only sell TDRs when a demonstrated demand exists that is not being met by the private market. The sales price for its TDRs can be set either by the bank board or by the sale of credits via sealed bids with the goal of maximizing the value of its TDRs.

Because the Chesterfield program eliminates the need for discretionary decisions, developers readily participated knowing that they would be able to complete their projects without delay, modifications, or additional last-minute costs. By 2007, over 90 percent of the units planned for Old York Village were either occupied, under construction or in the approval process. All residential and commercial development in Old York results in preserved farmland. The program has gained nationwide recognition as a model for combining smart growth and farmland preservation and won the American Planning Association's Outstanding Planning Program Award.

Chestnuthill Township, Monroe County, Pennsylvania, population 16,906 (2018), uses TDR to save farms, parkland and natural areas. TDR offers special development standards as well as density bonus to projects in receiving zones.

Chico, California, population 94,529 (2019), allows transfers from environmentally significant sending sites to more suitable receiving sites using the planned unit development, specific plan, or development agreement process.

Chisago County, Minnesota, population 56,579 (2019), allows transfers between properties in the Development Transfer District in order to protect farms and natural resources in its Green Corridor and St. Croix River overlays.

Churchill County, Nevada, population 24,909 (2019), uses TDR to preserve agricultural and environmental sending sites surrounding Fallon Naval Air Station (NAS). Bonus TDRs are allocated to sending sites that provide water recharge, public access and military installation buffer benefits. TDRs are transferred to residential, commercial, and industrial planned unit developments in urbanizing areas further than three miles from Fallon NAS.

Clallum County, Washington, population 77,331 (2019), uses TDR to protect farmland and open space with transfers to residential receiving zones.

Clarkdale, Arizona, population 4,271 (2019), adopted a TDR ordinance in 2017 to promote the preservation of open space, natural areas, and historic resources; the installation of infrastructure and public art; and the reduction of water use.

Clatsop County, Oregon, population 40,224 (2019), promotes density transfer to preserve lands suitable for permanent open space in the North Clatsop Plains.

Clearwater, Florida, population 115,159 (2019), uses TDR to implement goals for revitalization and protection of environmental, archeological, historical, and architectural resources. Rights are transferred at a 1:1 ratio and can be converted from one land use to another using trip generation rates. During a six-year period, 180 units transferred, primarily between beachfront sites.

Clifton Park, New York, population 36,566 (2018) uses Open Space Incentive Zoning to preserve farms, trails, and natural resources. On receiving sites, developers can gain one additional single-family residential unit by saving three acres of sending area land or paying $30,000 to the town's open space fund. Similar options are available for duplex units and office or retail floor area.

Collier County, Florida, population 384,902 (2019), surrounds the City of Naples in the southwestern corner of Florida and includes portions of the Florida Everglades as well as numerous islands along its Gulf of Mexico shoreline. After decades of environmental damage caused by clear-cutting and ill-conceived real estate deals, Collier County has become increasingly environment friendly. Today, federal, state, private non-profit, and county efforts have preserved roughly 80 percent of Collier County in parks and preserves that include the Audubon Society's Corkscrew Swamp Sanctuary, Fakahatchee Strand Preserve, the Florida Panther National Wildlife Refuge, and Big Cypress National Preserve. The county's preservation efforts have included three different TDR programs.

Special Treatment Overlay - In 1974, Collier County created the Special Treatment (ST) Overlay zoning designation to protect coastal areas, wetlands, habitat, and other places of ecological significance. The ST overlay instituted strict environmental regulations and required approval of a special permit for all new development. As an alternative to on-site development, property owners were encouraged to preserve sensitive land using TDR. However, by 1981, the ST TDR mechanism had transferred only 526 development rights, protecting 325 acres, with most of that activity occurring in a single transaction of 350 TDRs. The relatively low rate of use was attributed to the fact that developers were often able to build what they wanted on site by concentrating development on the non-sensitive portion of sites using the planned unit development process and using the sensitive ST land as an open space amenity.

According to the Collier County Land Development Code Sec. 2.03.07D as it appeared in January 2021, the number of transferrable development rights is based on the density allowed by the underlying zoning of a sending site. When TDRS are transferred from urban areas to urban areas, TDRs can be transferred into all residential zoning districts and the residential components of planned unit developments. The maximum density allowed by the zoning of the receiving site serves as baseline and the bonus density is ten percent of baseline in the lower density residential zones (up to RMF-12) and five percent in the RMF-16, RT and PUD districts. Once the Board of County Commissioners approves these urban-to-urban transfers, the owner must dedicate the protected ST land to a county, state, or federal agency or a private non-profit conservation organization. A 2007 study from Resources for the Future reported that this requirement has reduced enthusiasm to use this mechanism.

The TDR mechanism adopted for the ST Overlay in 1974 is now intertwined with the transfer process available in Collier County's Rural Fringe Mixed Use (RFMU) District (discussed below). According to the code as it existed in 2017, the rules for RFMU transfers are more welcoming than those that apply outside of the RFMU. Sending areas in the RFMU are allowed one on-site dwelling per 40 acres, but owners can transfer one TDR per five acres and sending site owners are not required to convey title when a transfer is approved. In addition, various bonus TDRs are available to sending site owners as explained in the RFMU district description below.

Rural Lands Stewardship (RLS) – The 300-square mile area surrounding the City of Immokalee, 30 miles northeast of Naples, is zoned for agriculture but contains many environmentally-significant resources including wetlands, flow-ways, water retention areas, aquifer recharge zones, and critical habitat. To head off protracted litigation, the major landowners came to an agreement with environmentalists and government officials in 1999 that produced a Rural Land Stewardship (RLS) program aimed at preserving the most sensitive areas and transferring the unused development potential to a new town located on less-sensitive land.

The program has a 195,000-acre planning area with the majority of the land under a few large ownerships. The agricultural zoning allows one on-site dwelling per five acres. The sending sites are proposed by the property owners and the stewardship credit allocation is calculated by a two-phase process. First, the land itself is assigned preservation value according to the priority status of its environmental resources, meaning habitat, flow-ways, or aquifer recharge zones. In the second step, the applicable property owners choose the extent to which they want to restrict the use of their property. In a tiering of restrictions, owners can decide to only prohibit residential development or also restrict other activities such as mining, recreation, cropland, agricultural support, and pasturing. The amount of credits available to a property grows as the owner chooses to apply greater restrictions.

The new town of Ave Maria is the receiving area for this RLS program. Here, one acre of receiving area land can be developed to the maximum extent allowed under the Ave Maria master plan when a developer retires eight credits. With this approach, the RLS program avoids the problem inherent in many TDR programs in which developers must retire TDRs to reach a maximum density that may also be the density preferred by the jurisdiction. The possibility of giving more development potential to one acre of receiving area versus another was not a problem in Ave Maria because a single partnership is developing the entire town.

To reduce the need to commute back and forth to Naples, Ave Maria aims to be a self-supporting, mixed-use community planned for 11,000 residential units, 1.7 million square feet of retail, office, and business park structures plus a variety of recreational facilities. As of 2020, the estimated population was 10,000, including enrollment at Ave Maria Catholic University, which alone is expected to ultimately accommodate 5,000 students. Developers are assured of their ability to use RLS credits and achieve the densities approved in the Ave Maria plan. Furthermore, there are no mechanisms allowing circumvention of the TDR requirements.

To date, Collier County's RLS has placed 54,962 acres of sending area land under easement, which is more gross acreage preservation than the amount achieved by many US TDR programs. The mere existence of Ave Maria was facilitated by the fact that the receiving area is surrounded by its own undeveloped sending area and consequently devoid of nearby residents that might be inclined to oppose a relatively compact new town development next door. However, critics observe that the remote location of Ave Maria conflicts with basic principles of growth management (Linkous & Chapin 2014). Another study argues that the method of sending site selection in this program does not result in an optimal outcome for sensitive environmental resources including panther habitat (Schwartz 2013).

Rural Fringe TDR Program - Collier County adopted a third TDR program in 2004 for its Rural Fringe Mixed Use (RFMU) district, a 73,222-acre area owned by approximately 10,000 different property owners. The Rural Fringe TDR program aims to protect sending lands containing large areas of wetlands, endangered species habitat and other significant environmental resources. The sending and receiving areas are mapped. The sending area contains 21,128 acres of land under private ownership and 20,407 acres held by public entities. The receiving area contains 22,020 acres. The RFMUD also has a neutral area with 9,667 acres.

In the sending areas, on site development is allowed at a density of one unit per 40 acres or one unit for every lot smaller than 40 acres created before 1999. Alternatively, sending area property owners can receive one Base TDR Credit per five acres placed under a qualifying easement. The county has set a minimum price of $25,000 per Base TDR Credit but bonus credits can be sold at any price negotiated between buyers and sellers.

For each Base Credit, sending site owners can gain one early entry bonus credit, an incentive designed to jump start the program which was still in existence as of 2017. Sending area owners can also gain one Environmental Restoration and Maintenance Bonus TDR Credit when they elect to submit and implement an approved Restoration and Maintenance Plan. These plans must be prepared by a county-approved environmental contractor and are required to include a listed species management plan, a procedure to remove invasive vegetation, and financial assurance that the plan will be followed until the property has achieved sustainable ecological functionality or the property is conveyed to a public agency.

For each Base Credit, sending area property owners can obtain a third bonus credit known as a Conveyance Bonus TDR by conveying title in fee at no cost for a property with an approved Restoration and Maintenance Plan to an appropriate governmental agency. By allowing sending sites to gain all four levels of TDR credits, Collier County has created meaningful incentives for sending site property owners to choose preservation/restoration rather than on site development. As an example, the owner of a 40-acre sending site has the choice of building one dwelling unit on site or selling as many as 32 TDR credits.

Despite these seemingly generous allocation mechanisms, Collier County was not satisfied with the rate of preservation when it looked at various ways of retooling the RFMU transfer program in 2016 and 2017. Although staff saw sufficient demand for TDR credits in the long-term future, the RFMU has enough entitled lots to accommodate near term demand. Developers could consequently decline to use TDR by continuing to build relatively low-density, clustered projects. Staff proposed doubling the allocation rate to eight TDC credits per five acres in an effort to incentivize development at higher, smart-growth densities. Another option floated in 2017 involved the establishment of a capitalized TDR bank capable of buying a significant amount of TDR credits and holding them until the anticipated long-term demand for TDRs materialized. In 2017, the county was also wrestling with the fact that conservation agencies at that time were reluctant to accept ownership to properties from many sending area property owners willing to donate them after selling their TDR credits.

Despite the large inventory of entitled properties and the difficulties of finding a conservation agency willing and able to own and manage sending sites, Collier County's RFMU program had 7,347 acres of sending area land under recorded development right limitations as of June 2019, which represented a 12 percent increase in preserved land over a three-year period. This 2019 preservation snapshot reflected the recording of 1,491 Base TDR Credits, 1,491 Early Entry Credits, 954 Restoration and Maintenance Credits, and 432 Conveyance Credits.

References

Linkous, E. & T. Chapin. 2014. TDR Performance in Florida. Journal of the American Planning Association. Volume 80. Issue 3. Pages 253-267.

Schwartz, K. 2013. Panther Politics: Neo-liberalizing Nature in Southwest Florida. Environment and Planning. Volume 45, pages 2323-2343.

Conewago Township, Adams County, Pennsylvania, population 7,220 (2018), uses TDR to redirect growth potential from sending areas with farmland, natural resources, and rural character into a 150-acre receiving zone with a 10 percent coverage baseline where each TDR yields 1,000 more square feet of coverage.

Coral Gables, Florida, population 50,226 (2019), uses TDR to encourage the preservation of historic landmarks and the expansion of public open space with transfers allowing up to 25 percent additional floor area on receiving sites.

Cottage Grove (Town), Dane County, Wisconsin, population 7,106 (2018), replaced an earlier TDR code by opting into the Dane County TDR program in 2011 and rezoning much of its Agricultural Preservation Area as a TDR Sending Overlay Zone. Receiving sites can build eight additional dwelling units per TDR.

Crested Butte, Colorado, population 1,339 (2019), incorporates transfer provisions for land annexed into the town. One additional dwelling unit or 5,000 square feet of commercial floor area is allowed for the preservation of five acres in Priority Preservation Areas, including ecosystems and cultural sites, or three acres in Hazard Areas prone to avalanche, flooding, landslide and inaccessibility.

Crystal River, Florida, population 3,129 (2019), adopted TDR to preserve Three Sisters Spring. That sending site is now saved but TDR remains in the city code.

Cupertino, California, population 59,276 (2019), sits at the eastern base of the Santa Cruz Mountains, ten miles west of downtown San Jose. It is home to the Headquarters of Apple and in 1973 it adopted a form of TDR that allowed Apple to continue thriving within the city.

As the tech industry was starting to boom in California's Silicon Valley, Cupertino realized that traffic would need to be managed in order to maintain acceptable levels of service in its DeAnza/Stevens Creek commercial corridor. If left unchecked, unrestricted development in this corridor would require roadway widenings that were considered unfeasible. To implement the Cupertino General Plan's policy on this issue, the city adopted its Traffic Intensity Performance Standard (TIPS), which established a baseline level of development in the DeAnza/Stevens Creek Corridor in terms of afternoon peak-hour motor vehicle trip generation. Specifically, the baseline was 16 one-way trips per acre in the PM peak hour. This baseline could be converted to various types of land uses using PM peak hour trip generation factors contained in the city's Development Intensity Manual: 0.75 trips per residential unit, 1.0 trip per 1,000 square feet of office space, 2.0 trips per 1,000 square feet of general retail floor area, 3.5 trips per 1,000 square feet of restaurant and so forth. The trip generation rates for land uses not listed in the TIPS Manual could be calculated based on a traffic engineering study or preexisting research findings.

Cupertino realized that some developments would not reach the baseline of 16 PM peak-hour trips per acre while others would need to exceed that baseline in order to locate within the corridor. Consequently, TIPS allowed PM peak-hour trip rights to be transferred from sending to receiving sites within the DeAnza/Stevens Creek planning area by conditional use permit (CUP). In addition to the usual findings required to approve a CUP, the city had to determine that the transfer would retain a viable level of development on the sending site. Covenants are recorded on the sending and receiving sites to document these transfers after approval.

At its peak, trip rights in this program were selling for $50,000 each. Some developers acquired trip rights before they needed them in anticipation of rising value over time. After 40 transfers, development had reached the capacity for the entire corridor. Apple's 750,000 square foot research and development office park was made possible by the acquisition of 322 trip rights from three separate sending sites. Consequently, this program allowed a major employer to stay in Cupertino without overwhelming the capacity of the city's transportation network. Cupertino provided a model for using transfers to allow necessary flexibility in the location of development within finite infrastructure systems. This model has since been emulated in other California cities including Burbank, Irvine, and El Segundo.

Dallas, Texas, population 1,343,573 (2019), allows administratively approved 1:1 transfers from historic properties that permit receiving sites up to FAR 4 of additional floor area. But high baseline density hampers the need to use TDR.

Dane County, Wisconsin, population 546,695 (2019), adopted a TDR code that towns have the option of using for farmland preservation and growth control.

Dane (Town), Dane County, Wisconsin, population 1,137, (2018), uses the Dane County TDR overlay zones to motivate the preservation of farmland by transferring development potential to receiving sites less suitable for agriculture.

Delray Beach, Florida, population 68,217 (2019), encourages historic preservation by TDR. A TDR equals one unit or 2,000 square feet of floor area.

Denver, Colorado, population 705,576 (2019), promotes historic preservation using transfer of undeveloped floor area, calculated as lot size plus the difference between the maximum floor area allowed by zoning and the floor area of the landmark plus four times the square feet of street-facing exterior of a rehabilitated landmark. TDR has preserved at least three Denver landmarks.

Douglas County, Colorado, population 351,154 (2019), promotes permanent open space preservation with density transfers that maintain net development.

Douglas County, Nevada, population 48,905 (2019), lies partly within the Sierra Nevada mountains at the California-Nevada border and extends east to include farmland in the Carson Valley. The westernmost portions of Douglas County are within the Lake Tahoe watershed and consequently under the jurisdiction of the TDR program managed by the Tahoe Regional Planning Agency. However, Douglas County also manages a separate TDR program aimed at preserving resource lands in its Agricultural designation (A-19) and its Forest and Range designation (FR-19).

Douglas County's original TDR program, launched in 1996, languished primarily because the TDR allocation ratio made TDRs too expensive for receiving site developers to afford. Under the initial formula, sending site owners could transfer two TDRs for every 19 acres of permanently preserved land. In addition, developers could originally gain density by using planned developments in the receiving areas, a loophole that the county later closed.

In 2001, Douglas County brought the program to life with amendments that increased the allocation ratio. In the A-19 zone, sending site allocation was changed to nine bonus TDRs per 19 acres plus seven additional TDRs for 19 acres when at least half of the 19 acres is located within the 100-year floodplain as mapped by the Federal Emergency Management Agency (FEMA). Sending site owners can transfer another seven TDRs per 19 acres by restricting the transfer of surface and groundwater irrigation rights from the sending parcel. The county board can add another TDR per 19 acres when owners grant permanent public access easements to bodies of water, public lands, or historic sites. By foregoing on-site development, the sending site owner also gets to transfer the unused base unit. Using all of these options, a sending site owner can transfer 25 TDRs per 19 acres. Another 20 TDRs can be transferred per 100 acres from sending site parcels at least 100 acres in size.

There are fewer allocation options in the FR-19 zone. Here, one bonus TDR can be transferred per 19 acres when at least half of the 19 acres is located in the 100-year flood plain. The county board can add another TDR per 19 acres when owners grant permanent public access easements to bodies of water, public lands, or historic sites. By foregoing on site development, the sending site owner also gets to transfer the unused base unit. Using all of these options, a sending site owner can transfer three TDRs per 19 acres. Another one TDR can be transferred per 100 acres from FR-19 sending sites at least 100 acres in size.

The receiving areas are mapped, and each TDR allows one dwelling unit above baseline density. The price of TDRs are negotiated between buyers and sellers.

The 2001 code changes produced a robust TDR market for the next eight years. In 2002 alone, over 1,262 TDRs were certified, representing 2,177 acres of deed restricted land. As of 2009, 4,003 acres had been placed under easement and 3,921 TDRs had been certified, of which all but 206 TDRs had been transferred. Of the total acreage preserved, 2,892 acres, or 73 percent, are in floodplains, an indication that the allocation formula is implementing the goal of Douglas County's Living River policy of retaining rivers in their natural state and allowing water to access the floodplain.

Since 2009, the program has seen very little use and the county has recognized the need to make further changes. As early as 2007, Douglas County began thinking about various modifications to its TDR program. Consideration was given to allowing properties not zoned A-19 or FR-19 to become sending sites if they are located on hillsides or within floodplains. The 2011 Master Plan also noted that all 4,003 acres preserved as of 2009 were located in the Carson Valley watershed and that the market could be improved with a master plan amendment allowing TDRs to be transferred between the Carson Valley watershed and the Topaz watershed. At that time there was no demand for TDRs from the Topaz watershed because the 1,322-acre receiving area was not yet under development. Master plan drafts since then have called for consideration of requiring TDRs for proposals requiring rezonings and the possible formation of a Douglas County TDR bank.

A 2020 draft of the county's forthcoming master plan observes that outside the 3,705 acres of undeveloped receiving area land, developers do not have to buy TDRs when their rezonings are approved. The 2020 draft allows readers to draw the logical conclusion that developers are avoiding the receiving areas and using upzonings outside the receiving areas to gain bonus density for free rather than paying for it. The draft then repeats the 2011 master plan by calling for consideration of requiring TDRs for all of the additional units gained by rezoning requests as well as the possibility of forming a county TDR bank.

Douglass Township, Berks County, Pennsylvania, population 3,599 (2018), preserves farms, nature, and open space with transfers allowing age-restricted projects to exceed density, height, coverage and impervious surface baselines.

Dover, New Hampshire, population 31,577 (2019), adopted an industrial TDR program in 1990. Based on its success, the city expanded that program and added a residential TDR program in 2003 aimed at preserving significant conservation features including farmland, wetlands, habitat and landmarks.

East Hampton, New York, population 22,009 (2016), uses a TDR program to facilitate the creation of affordable housing while protecting land with natural, scenic, recreational, forest, historic, aesthetic, cultural, or economic value.

East Nantmeal Township, Chester County, Pennsylvania, population 1,846 (2018), uses TDR to preserve prime farmland with a 5:1 transfer ratio. Owners often prefer the cluster option and purchase of development rights alternative.

East Vincent Township, Chester County, Pennsylvania, population 7,327 (2018), uses TDR to preserve farms, nature and rural character. Receiving site bonus development ranges from 1.25 single-family dwellings per TDR to 2.5 mobile homes per TDR. Projects using five or more TDRs can also use TDRs to increase impervious lot coverage or add floor area above baseline height limits.

Easthampton, Massachusetts, population 15,829 (2019) uses TDR to preserve farmland, aquifer recharge areas, rural character and historic resources. Receiving site developers can use one TDR or a cash contribution of $35,000 to gain specified amounts of density, non-residential floor area or one neighborhood commercial building within a Traditional Neighborhood Development.

Eden, New York, population 7,670 (2017), adopted TDR in 1977 to preserve agricultural and environmental areas. In 2000, the town eliminated a sending area that offered a 30-to-one transfer ratio as an incentive for participating. The use of TDR is not subject to discretionary approval. The program has produced a few transfers including one that added density to a senior citizen development.

El Paso, Texas, population 679,813 (2019), aims to preserve open space, archeological sites and natural resources by allowing the transfer of development rights to receiving areas in six zoning districts at a one-to-one ratio.

El Segundo, California, population 16,731 (2020), uses transfer of floor area, with no net gain, between properties in common ownership primarily to improve traffic circulation but also to promote open space and other community benefits.

Everett, Washington, population 109,766 (2019), offers administrative approval of transfers from land precluded from reasonable use by environmental rules.

Exeter, Rhode Island, population 6,561 (2018), uses TDR to preserve farmland, open space, historic sites and natural resources using two programs. An interjurisdictional program pairs Exeter sending areas with receiving areas in adjacent North Kingston. Upon approval of a Planned Village Overlay District, TDR is required to exceed baselines determined by the maximum density of the underlying zoning. The second program establishes three categories of TDRs depending on the preservation values of each specific sending site. Higher value TDRs allow higher amounts of bonus single family, townhome and multi-family units. Developers may also use in lieu fees established by the appraised land value of a parcel considered most consistent with the town's preservation goals. Developers pay in lieu fees based on the revenue stream from lot or home sales. In lieu revenues are reserved for farmland or open space preservation.

Falmouth, Massachusetts, population 31,531 (2010), uses TDR to protect coastal ponds and groundwater recharge areas in an effort to protect water quality. The density bonus allowed on receiving sites varies depending on the zoning of the donor site. The program has been used at least three times, primarily preserving water resource protection areas including nine acres of woods near Long Pond, which is the primary source of the Town's drinking water.

Farmington, Utah, population 25,339 (2019), adopted a TDR program in 2018 allowing receiving sites to reduce on-site open space and effectively increase density by transferring TDRs from open space sending sites. As of 2019, the city bought land for a park and sold 35 of its 100 TDRs at prices ranging from $3,500 to $85,000. All proceeds must be used exclusively to buy or improve park land.

Framingham, Massachusetts, population 72,308 (2019), uses TDR to preserve farms and open space. TDR can increase receiving site floor area, height and lot coverage.

Frederick County, Virginia, population 89,313 (2019), uses TDR to preserve farms, forest lands, natural resources, open space and scenic views. As of a 2013 revision, transfer ratios apply to TDRs from priority sending areas as well as added bonus ratios for attached and multiple-family units on receiving sites. As of May 2019, at least nine TDR sending site applications had been approved.

Fremont County, Idaho, population 13,099 (2019), adopted a mechanism in 1992 allowing cluster developments to gain bonus density by preserving offsite farmland and environmental areas. Four transfers saved 160 acres but the tool was removed from the county's development code in 2011.

Ft. Lauderdale, Florida, population 180,124 (2019) allows density transfers from the Intercoastal Overlook Area to the North Beach Residential Area in order to protect views from and to the Intercoastal Waterway.

Ft. Myers, Florida, population 79,927 (2019), has a bonus incentive program in the smart code that regulates the city center. Developers can gain one additional dwelling unit per payment of a $10,000 community contribution that the city uses for affordable/workforce housing, public open space/recreational areas, infrastructure, public transit, public parking and other downtown facilities.

Gallatin County, Montana, population 114,434 (2019), uses TDR in five of its 22 zoning districts including the Middle Cottonwood, where at least 15 dwelling units were transferred via planned unit development, preserving 514 acres of habitat.

Gilpin County, Colorado, population 6,243 (2019), uses TDR to preserve sending areas that are highly visible, accessible or near year-round waterways.

Gorham, Maine, population 17,651 (2018), charges a development transfer fee of $15,000 per bonus dwelling unit in receiving zones without offering the option of transferring actual TDRs. The total fee for bonus units in an entire receiving site development is divided by all units in the project and each unit is charged prior to building permit issuance. The town uses transfer fee proceeds to buy open space, historic sites, farms, habitat, and land with other natural resources.

Greenville County, South Carolina, population 523,542 (2019), adopted a TDR ordinance in 1983 as a means of mitigating the downzoning of the Paris Mountain planning area in order to maintain a level of development that could be accommodated by the area's only access road. Owners of land with natural development constraints and marginal views were motivated to transfer development potential to land with higher value view lots. The county reduced the uncertainty often associated with TDR programs by dispensing with hearings or discretionary decisions and making the entire process administrative. As a result, at least ten transfers have been approved.

Groton, Massachusetts, population 11,386 (2018), adopted TDR provisions in 1980 to protect scenic areas, environmentally sensitive land, farmland, and land important to the town's water supply. In 1988, Groton allowed developers to use TDRs either to increase receiving site density or to build up to six extra dwelling units per year in addition to the six dwelling units per subdivision per year that would otherwise be permitted under the town's development permit limit provisions. Between 1988 and 2002, TDR preserved an estimated 500 acres of land including farmland, land important to Groton's water supply and the Nashua River Greenway. During that time, developers used TDRs exclusively to exceed the permit quotas rather than increase density and TDRs were used in every subdivision with more than six undeveloped lots.

Gunnison County, Colorado, population 16,802 (2019), is located in central Colorado, 200 miles southwest of Denver. Almost 80 percent of the county's land area is managed by federal agencies including the Bureau of Land Management, the National Forest Service, and the National Park Service. The remainder of the county consists of ranch land, the City of Gunnison, and four towns including Crested Butte, a former mining town now in the National Register of Historic Places, and Mount Crested Butte, home to a Crested Butte Ski Resort.

The voters of Gunnison County approved a ballot measure in 1997 creating a Land Preservation Fund capitalized by sales tax revenues. Using $3.6 million of this dedicated sales tax income, the county leveraged $44 million as of 2021 for preservation by matching its money with non-county sources, particularly Colorado's state-run lottery revenues. This resulted in an average cost to the county's Land Preservation Fund of $355 per preserved acre. As of 2021, Gunnison County had preserved over 16,000 acres, or roughly four percent of the county's private land.

Gunnison County adopted a Residential Density Transfer (RDT) program in 2009 allowing developers to increase the density of their subdivisions when they contribute to the preservation of offsite ranchland. Developers are awarded this extra density by paying a density transfer charge calculated as ten percent of the land value increase generated by the extra density.

A qualifying receiving site project in Gunnison County's RDT program can include any residential subdivision with five or more residential lots or residences. Participating developers can reduce their on-site open space requirement from 30 to 15 percent of the total project area and increase the density of the receiving site by using RDT. The Gunnison County Assessor uses mass appraisal techniques to estimate the land value of the project site before and after subdivision approval. Developers are then required to pay ten percent of this increase in value. Mobile homes and lots used exclusively for workforce housing are excluded from this calculation. This calculation is only performed for the residential portion of a mixed-use development.

An agreement between the receiving site developer and the county enforces compliance with RDT payment requirements. This agreement is executed prior to the recording of the final plat. Developers can pay some or all of the required RDT payment upon final plat recordation. A ten-percent discount applies to all payments made prior to final plat recordation. Alternatively, developers can choose to make these payments as individual lots are sold.

The county adds the revenues from RDT payments to the Land Preservation Fund to augment the revenue from the open space sales tax. As mentioned above, the county uses the Land Preservation Fund to leverage the Colorado lottery and other funding sources exclusively to purchase land and conservation easements that permanently preserve parcels with significance as open space, agriculture, habitat, wetlands or watershed protection.

Gunnison County's RDT program resolves several issues that can derail traditional TDR programs. Sending areas are not designated and a TDR allocation formula is not contained in the RDT ordinance, thereby avoiding disputes that sometimes arise over who can sell TDRs and how many TDRs will be issued in return for preserving a sending site. Essentially the sending site is selected by the county when it decides where to spend the Land Preservation Fund dollars.

Likewise, an economic analysis was not necessary to determine how many TDRs would be needed per unit of bonus density because TDRs are not involved. The density transfer charge did not need to be legislated because compliance is calculated by assessment of the value added by the additional development for a specific receiving site. This eliminates another potential source of disagreement that can hamstring adoption of traditional TDR ordinances.

The RDT program motivates developer participation by allowing much of the RDT payment to be delayed until the developer has money in hand from the sale of lots. Developers do not have to negotiate and buy TDRs from sending site owners because there are no TDRs, and the county performs the sending site preservation. In addition, because RDT compliance is in the form of a cash payment, the county is relieved of having to create, track and retire actual TDRs. Using cash also allows the county to target the best acquisitions and gain maximum leverage from RDT revenues.

A detailed analysis of this program, written by Mike Pelletier, Chris Duerksen, and Rick Pruetz, appears in a 2010 issue of the American Planning Association's *Planning Advisory Service Memo* entitled TDR-Less TDR Revisited: Transfer of Development Rights Innovations and Gunnison County's Residential Density Transfer Program.

Hadley, Massachusetts, population 5,346 (2018), uses TDR to protect farmland, offer fair economic return to farmland owners and promote compact development in areas served by public infrastructure. Restrictions placed on sending sites are permanent unless the site is no longer deemed suitable for agriculture and this conclusion is affirmed by a two-thirds vote of both branches of the Massachusetts general court. If the court rules that the sending site is no longer viable farmland, the landowner must buy back the development rights at their then fair market value.

On receiving sites, each TDR allows an extra 2,000 square feet of commercial or industrial floor area plus the possibility of reducing lot coverage from a maximum of 30 percent to up to 70 percent and the ability to reduce the required parking area from 2.0 times to 1.5 times the building's floor area. Alternatively, developers may use one TDR to add two additional bedrooms to a senior housing development. Receiving area developers can choose to comply using actual TDRs or by paying a cash contribution which Hadley then uses to buy agricultural easements. The payment is the per-acre cost of actual farmland easements over the last three years, as determined by the Conservation Commission, multiplied by the number of acres needed to satisfy the TDR requirements. The Hadley TDR program added $338,772 to the Land Preservation Fund which, with state funds, preserved 239 acres of farmland.

Harford County, Maryland, population 255,441 (2019), allocates one TDR per ten acres of preserved farmland and allows transfers to receiving sites within one-half mile of sending sites.

Hatfield, Massachusetts, population 3,284 (2018), uses TDR to preserve farmland, open space, historic resources and rural character. Developers can use the TDR mechanism to increase non-residential floor area and lot coverage. However, rather than buy and retire actual TDRs, receiving site developers comply exclusively by making a cash contribution to the Town's Land Preservation Fund which is used by the Hatfield Agricultural Advisory Committee to buy easements in the sending area. Under the Hatfield formula, the average per acre value of a development right is the difference between the average per acre assessed value of residentially improved land and the average assessed value of unimproved land according to the most recent Town-wide comprehensive property value assessment. The cash contribution in lieu of each required TDR must be one and one half times the average per acre cost of a development right as determined by the above formula.

Hebron (Town), Connecticut, population 9,482 (2018), adopted TDR to redirect development away from Amston Lake to limit environmental impact on the lake as well as reduce congestion and depletion of groundwater supplies in the area.

Hellam Township, York County, Pennsylvania, population 6,043 (2010), uses TDR to preserve farmland and open space. TDRs can be used to exceed baseline density in residential receiving areas or to exceed impervious lot coverage baselines in non-residential as well as residential receiving areas.

Hereford Township, Berks County, Pennsylvania, population 2,975 (2018), uses TDR to preserve farmland, sensitive natural areas and rural community character. TDRs can be transferred back to a property from which they have been severed as long as this reattachment does not give the site more TDRs than it had originally and as long as the TDRs come from the same zoning district. Receiving site projects can use TDR to increase residential density, lot coverage and floor area in excess of baseline building heights.

Highlands County, Florida, population 106,221 (2019), uses TDR to preserve wetlands, certain uplands and other important land identified by the county board. Participants must donate environmentally sensitive sending sites to the county in order to be issued development rights certificates. Transfers of less than 20 units can be approved administratively. Transfers of 20 or more units must be approved by the Board of County Commissioners. The program had protected 286 acres as of 2014.

Hillsborough County, Florida, population 1,472,000 (2019), promotes 1:1 transfers to preserve farmland, environmental areas and historic resources.

Hillsborough Township, Somerset County, New Jersey, population 38,303 (2010), has offered TDR since 1975 as a way to preserve farmland, protect environmentally sensitive land, secure sites for public facilities and reduce the cost of streets and other infrastructure. Development credits are issued to sending sites based on the density allowed by zoning and can be transferred to five receiving zones. The program produced at least one transfer, resulting in the township accepting title to a 70-acre farm.

Hollywood, Florida, population 152,511 (2019), uses TDR to reduce beachfront density to protect environmental resources, gain coastal access, and keep buildout within the planned capacity for the sewer, roadway and utility systems.

Honey Brook Township, Chester County, Pennsylvania, population 8,311 (2018), uses TDR to preserve farmland, watersheds, natural areas, and community character. Receiving sites using TDR to increase residential density can boost the bonus by 50 percent for units restricted to persons over 55 years of age or workforce units. TDRs can also be used to exceed baselines for impervious surface coverage and floor area in excess of baseline building height. At least 29 TDRs have transferred, preserving 58 acres.

Hopewell Township, Mercer County, New Jersey, population 17,725 (2019), uses TDR to preserve farmland and conservation areas by redirecting growth to receiving sites via non-contiguous cluster development (Section 17-172j).

Hopewell Township, York County, Pennsylvania, population 5,566 (2018), adopted a TDR program to preserve farmland and conservation areas which has been used at least once.

Howard County, Maryland, population 325,690 (2019), is located 20 miles southwest of Baltimore and 30 miles north of Washington, DC. The County has a Density Exchange Option aimed at preserving farmland.

As of 2021, the Density Exchange Option (DEO), Section 106 of the Zoning Code, allows sending sites to be approved on parcels within the DEO Overlay District zoned Rural Conservation (RC) that are capable of accepting a conservation easement at least 20 acres in size. One development right is created per three acres preserved. Although a parcel zoned RC can be either a sending or receiving parcel, a single property cannot be both a sending and receiving site.

Receiving sites are also within the DEO Overlay and zoned either RC or Rural Residential (RR). However, receiving sites zoned RC must be less than 50 acres in size and be adjacent to properties no greater than 10 acres in size on 60 percent of their perimeter.

In a variation known as Cluster Exchange Option (CEO) one development right can be created per 4.25 acres preserved and receiving sites must be within the DEO Overlay and zoned RC.

On both the DEO and CEO receiving sites, baseline density is one dwelling per 4.25 acres. Each development right allows one bonus unit on a receiving site up to a maximum density of one unit per two acres. Additional rules apply for certain minor subdivisions.

Receiving sites must be designed to cluster development in ways that maximize preserved farmland, minimize conflicts with farming operations, and protect rural and scenic quality, especially as seen from public roads. Receiving site developers can use TDRs by right, a feature found in many successful TDR programs because it relieves developers of the uncertainty, cost, and delay often associated with discretionary processes. As of 2016, the Howard County program had protected 4,980 acres (Maryland 2016).

Reference
Maryland. 2016. Transfer of Development Rights Committee Report. Maryland Department of Planning. Accessed 1-29-21 at TDR-committee-report-2016.pdf (maryland.gov).

Huntington (Town), Suffolk County, New York, population 203,264 (2010), motivates the preservation of natural open space using a Transfer of Density Flow Rights mechanism in which unused sanitary sewer flow credits from a sending site can be transferred to appropriate receiving sites needing additional credits for their proposed land uses and densities/intensities. Code section 198.118.2 has another TDR program allowing the Town to approve applications for transfers of traditional development rights that preserve sending areas with agricultural, scenic, natural, historic, cultural, aesthetic or open space value.

Indian River County, Florida, population 159,923 (2019), uses TDR in non-contiguous planned developments to promote environmental preservation and affordable housing. Despite strict limits for on-site development, TDR had only preserved 116 acres as of 2014 due to low demand to exceed baseline density.

Iowa City, Iowa, population 75,130 (2019), adopted form-based standards for its Riverfront Crossings District that include height transfers for historic preservation, open space, public right-of-way, and public art. The transferable floor area for a landmark is the total lot size multiplied by the number of floors allowed by code. Participating landmarks are protected not by easements but by a code section prohibiting demolition by neglect.

Irvine, California, population 287,401 (2019), uses a form of TDR that allows flexibility in the location of development in an area where baseline development potential is limited by PM peak hour vehicle trips in order to maintain traffic volume within transportation system capacity. Baseline is the number of PM peak hour trips generated by a 0.25 FAR general office. Actual baseline may be higher or lower than FAR 0.25 depending on whether the land use of a proposed building has a higher or lower PM peak hour trip generation rate than the office rate. Several projects have used this option.

Islamorada, Florida, population 6,433 (2019), uses TDR to preserve environmentally sensitive and flood-prone lands by transferring residential units and non-residential floor area to less-environmentally-sensitive receiving sites in numerous zones. Islamorada's Building Permit Allocation System does not apply to some new construction involving transferred rights. After floor area has been transferred, non-residential sending sites can still be used for affordable housing.

Island County, Washington, population 85,141 (2019), adopted a TDR program in 1984 to mitigate a downzoning of rural land designed to protect farms, forests and other natural resources. The program was replaced in 1998 by one that allows increased densities on receiving sites when developers redeem Earned Development Units granted to the owner of a farm or forest after approval of a farm or forest management plan. Between 1984 and 1995, the original program certified 149 TDRs, aided by transfer ratios as high as 20:1. In those ten years, only three receiving sites used TDRs despite the ability of developers to achieve substantial density increases in some circumstances using TDR.

Islip, New York, population 330,543 (2018), adopted a historic preservation TDR mechanism in its Planned Landmark Preservation Overlay Ordinance of 1975.

Issaquah, Washington, population 37,965 (2019), uses TDR to protect potential park sites and critical environmental areas, particularly salmon habitat, riparian corridors, and floodways. In receiving areas, TDR can increase residential density, building height and impervious surface coverage. In addition to the purposes typically established for a TDR bank, the Issaquah City TDR bank can sell TDR certificates in advance of property acquisition and use the funds to implement TDR goals including the purchase of land for open space or parks. Issaquah and King County used the King County TDR program in March 2000 to preserve a 313-acre sending area in King County which allowed an additional 500,000 square feet of floor area in an office complex within the City of

Issaquah. Within two years of adopting its own TDR program in 2005, Issaquah had signed an interlocal agreement with King County, purchased two environmentally significant parcels, and stocked its TDR bank with the resulting 42 TDRs.

Jericho, Vermont, population 5,070 (2018), uses a non-contiguous planned unit development process to approve transfers that preserve open space, farmland, forests, and passive recreational sites as well as implement other town goals.

Kennett Township, Chester County, Pennsylvania, population 8,254 (2018) uses TDR to preserve farms, woodlands, open space and sensitive natural areas. TDRs can be used to double baseline density, reduce open space, or exceed baseline lot coverage, impervious surface coverage, and building height.

Kent County, Delaware, population 180,786 (2019), uses TDR to protect farmland, open space and other natural resources as well as discourage sprawl. TDR allocation varies depending on a sending site's location within three priority zones. TDRs can increase a baseline density of three units per acre to five units per acre in Secondary Receiving Areas and seven units per acre in Primary Receiving Areas. The TDR ordinance specifies regulations applicable to TDR receiving area developments including street layout, pedestrian amenities, landscaping, parks, community facilities, non-residential uses, building orientation and architecture. As of 2009, easements were placed on 180 acres.

Ketchum, Idaho, population 2,791 (2019), adopted TDR provisions in 2008 aimed at preserving historic/architectural resources and adding public parks.

King County, Washington, population 2,252,782 (2019), surrounds Seattle and its suburbs. It has earned a reputation for leadership in conservation, preserving over half of its total land area by combining federal and state holdings with many county initiatives including a TDR program that leads the nation in the amount of permanently protected land (Pruetz 2012).

King County's first TDR program was adopted in 1988 and produced only one transfer. King County launched a second-generation TDR program in 1998 that used a three-year pilot project to test the effectiveness of interjurisdictional transfers from county sending sites to receiving sites within the county's incorporated cities. To motivate cooperation, King County offered to fund amenities within participating cities. The county budgeted $1.5 million in 1999 to acquire TDRs for a TDR Bank and another $0.5 million for amenities in cities willing to adopt inter-jurisdictional receiving areas. In 1999, the City of Issaquah and King County entered into an interlocal agreement resulting in the first interjurisdictional TDR transfer in the Pacific Northwest. In this transaction, a developer bought 62 TDRs from a county sending site which were used to add 500,000 square feet of floor area to a Microsoft office complex in the city.

In 2000, King County and the City of Seattle signed an interlocal agreement in which TDRs from agricultural sending areas in the county could be used to gain a 30 percent height bonus for residential buildings in the Denny Triangle Urban Village, a downtown district in need of revitalization. Per the agreement, King County pledged to spend up to $500,000 on Denny Triangle amenities including green streets, pedestrian/bicycle improvements, transit facilities/incentives, open space, storm water management, or public art/street furniture.

In a 2009 agreement with the City of Bellevue, King County agreed to fund $750,000 in stream improvements as part of an interjurisdictional TDR mechanism.

In 2013, Seattle and King County signed an interlocal agreement that marked the first use of a Washington law allowing tax increment financing to fund infrastructure only in cities that adopt TDR receiving areas capable of accommodating that city's fair share of TDRs from sending areas in other jurisdictions of the Puget Sound Region. This is the only way that cities in the State of Washington can use tax increment financing. Per this agreement, Seattle creates a receiving area for 800 TDRs from sending sites under county jurisdiction and King County pledges to dedicate up to $15.7 million of additional property tax revenue to pay for open space and transportation improvements in Seattle's South Lake Union and Downtown districts. This agreement details how and when the King County payments will be used. For example, $2.9 million is dedicated to green street improvements in the first ten years of the agreement followed by $7.8 million for parks and a community center in years 11 through 20, and $5 million for transportation improvements in years 21 through 25.

The interlocal agreements illustrate how carefully King County and its cities calibrate the amount of development allowed per TDR in response to the real estate market of each receiving area. For example, under the first interlocal agreement with Seattle, each King County TDR produced an additional 2,000 square feet of floor area, a ratio that motivated the developers of the Aspira and Olive 8 towers in Denny Triangle to buy King County TDRs. Varying that formula based on local conditions, each King County TDR yields 1,333 square feet of additional floor area under the 2009 interlocal agreement with Bellevue. Each King County TDR generates 12 different variations of bonus square feet for the first 200 TDRs transferred under the 2013 agreement with Seattle. Normandy Park's 2012 ordinance allows 4,300 square feet of bonus floor area per King County TDR and Issaquah's 2013 ordinance yields 1,200 square feet of additional floor area per King County TDR.

To qualify as a sending site, properties must be within the R-1 Urban Separator, RA-2.5, RA-5, RA-10, A (agricultural) or F (forest) zone. In addition, to qualify, a sending site must provide at least one of five community benefits: agricultural potential, forestry potential, critical wildlife habitat, open space, or regional trail connectors/urban separators.

Generally, the allocation of TDRs is based on the number of dwelling units that can be built on site according to the size of the property (minus submerged land) and its zoning. For example, the owner of a 20-acre property zoned RA-5 can build/retain one house on site, record an easement permanently restricting the site to one unit per 20 acres, and sell three TDRs. However, bonus TDR allocations can be gained in certain sending areas such as rural forest focus areas, agricultural lands, and legal non-conforming lots.

Sending areas in the five zones classified as Rural have allocations ranging from one TDR per 2.5 acres in the RA-2.5 zone to one TDR per 80 acres in the F zone. Land in the Urban classification of R-1 can also qualify as a sending site and is allocated four TDRs per acre.

As discussed above, interlocal agreements have created interjurisdiction receiving areas in the cities of Seattle, Issaquah, Bellevue, Normandy Park, and Sammamish. In addition, land under King County jurisdiction can qualify to be receiving areas. In urban areas under King County jurisdiction, receiving sites can occur on land zoned R-4 through R-48, NB, CB, RB or O. TDRs from sending sites in a Rural Forest Focus Area can be transferred to rural areas zoned RA-2.5 if four conditions are met: the site has domestic Group A public water service; the site is within ¼ mile of an area where lots are mostly smaller than five acres; the receiving project will not adversely affect significant resources or environmentally sensitive areas; and the project will not involve the extension of public services and facilities in a way that creates or encourages a trend toward smaller lots. Receiving areas cannot be formed on Vashon Island or Maury Island, within the Rural Forest Focus Area, or within the Noise Remedy Area surrounding Sea-Tac Airport.

Receiving sites under county jurisdiction occur in 12 zones. At the low end of the density range, the RA-2.5 zone has a baseline of 0.2 units per acre and a maximum density of 0.4 units per acre. At the high end of the density range, the Regional Business and Office zones have a baseline of 36 units per acre and a maximum density of 48 units per acre.

In addition to additional development potential, TDR can be used to satisfy traffic concurrency regulations that apply in King County's rural areas, meaning outside the Urban Growth Area. Development cannot occur in some rural travelsheds because of inadequate transportation infrastructure. TDRs can be used to satisfy concurrency requirements if the sending and receiving sites are in the same travelshed because each TDR represents a permanent reduction of development potential and therefore demand for transportation improvements. The same TDR cannot be used for both traffic concurrency and increased development potential.

A Rural TDR can also be used to increase Accessory Dwelling Units from a baseline of 1,000 square feet of floor area to 1,500 square feet.

King County commissioned a study that estimated that a reduction of 272 metric tons of GHG emissions over 30 years could result from a single TDR transaction that transfers the potential to build one dwelling unit from a rural sending site to a receiving site in downtown Seattle. On that basis, this study calculated a reduction of 19,000 metric tons of GHG emissions may have resulted from the 2001 interlocal TDR agreement that severed 70 TDRs from rural King County sending sites (preserving portions of the watershed that is Seattle's main source of water) and transferred these TDRs into downtown Seattle's Denny Triangle (Williams-Derry & Cortes 2011).

Between 2000 and 2014, over 50 developers purchased 518 TDRs directly from private owners, for an average of five private transactions per year and 34 TDRs bought and sold annually. These private transactions exchanged a total of $7.13 million in TDR value. With 1,105 TDRs allocated to private sending site owners and almost 300 TDRs redeemed, some participants are holding TDRs, indicating speculation in the TDR market. Between 2000 and 2020, the average sale price was $22,371 for a Rural TDR and $8,888 for an Urban TDR. (Urban TDRs are only used in unincorporated areas of King County.) The highest prices for TDRs occurred in 2000 ($45,000 each) and dropped below $10,000 each in 2014. TDR activity increased between 1999 and 2007, then dropped to zero in 2008 and slowly increased as the county recovered from the Great Recession.

King County leads the nation in the acreage preserved by TDR partly because of its TDR Bank. The TDR Bank is tasked with three roles. 1) Facilitating private transactions by buying TDRs when sending site owners want to sell and holding them until receiving site developers want to buy them. 2) Acting as a revolving fund by reinvesting the proceeds of its TDR sales in continuing land protection. 3) Acquiring sending sites of special significance to incorporated cities as a means of motivating these cities to enter into interlocal TDR agreements with King County.

In addition to these three primary tasks, the Bank sells options to buy TDRs, allowing developers to secure a specific number of TDRs at a known price for the period of time needed until receiving site project approval. The Bank also offers to create extended purchase and sale agreements allowing developers to reduce risk and up-front costs. The Bank incentivizes cities to enter into interlocal TDR agreements with King County because the Bank can negotiate revenue sharing provisions that dedicate a portion of the proceeds from TDR Bank sales to the funding of infrastructure, parks, streetscapes and other amenities within the participating city's receiving area.

King County's TDR Bank sometime buys TDRs using its Conservation Future Tax. This tax is made possible by a State of Washington law allowing counties to dedicate a portion of property tax exclusively to preservation projects. The King County Conservation Futures Tax generally generates about $10 million annually for conservation projects. By using Conservation Futures Tax revenue to buy TDRs, King County converts what would otherwise be a one-time acquisition into an ongoing revolving fund for preservation. In its most dramatic example, King County used $22 million of Conservation Futures Tax proceeds to purchase 990 TDRs representing the preservation of 90,000-acre forest 25 miles east of Seattle.

The bank sells TDRs for a price considered affordable to receiving site developers. This price is lower than the cost of buying some of these TDRs. For example, King County bought TDRs from one high priority sending site for $171,333 each, while the average price paid by developers was $25,000. King County is able to do this because it has acquired some TDRs at prices below that average, such as TDRs purchased in the early years of the program. It may also be likely that King County is not fixated on recouping the cost of each TDR since the program's primary goal is preservation. Plus, any income from the sale of TDRs is more money than the county would receive if it only used the traditional approach of using money one time to buy land without selling the development rights or buying typical conservation easements as in a purchase of development rights program.

According to the King County website, the TDR program had protected over 144,500 acres between 1998 and 2019. King County included TDR in its Strategic Climate Action Plan (SCAP) as one of many implementation tools aimed at achieving its goal of reducing GHG emissions 50 percent by 2030 and 80 percent by 2050 compared with a 2007 baseline. The 2015 SCAP cites the county's TDR program for its multiple benefits of conserving forests, preserving farmland, and curbing urban sprawl, actions that mitigate GHG emissions and sequester carbon. In its report card on the 2015 SCAP, one of the top ten accomplishments was the Land Conservation Initiative, King County's accelerated protection of the best and last remaining open spaces, farmlands, forests, parks, and trails. TDR is recognized in the highlights of this Land Conservation Initiative. For example, interjurisdictional transfers between King County and the City of Sammamish preserved some of the envisioned greenbelt while protecting habitat for salmon and other critical species.

References

Pruetz, Rick. 2012. Lasting Value: Open Space Planning and Preservation Successes. Chicago: Planners Press, American Planning Association.

Williams-Derry, Clarke & Erik Cortes. 2011. Transfer of Development Rights: A Tool for Reducing Climate-Warming Emissions – Estimates for King County, Washington. Seattle: Sightline Institute.

Kitsap County, Washington, population 271,473 (2019), uses TDR to preserve rural character, farmland, and open space. Receiving areas are located within urban growth areas and designated in the comprehensive plan. In addition, the board of county commissioners may require TDRs for urban growth boundary expansions, site specific comprehensive plan amendments and rezonings. Development rights purchased for a site-specific amendment may also count towards any future rezone request within the new designation. By resolution, the county requires 1 TDR per rezoned acre and anywhere from one TDR to three TDRs per acre for comprehensive plan amendments depending on the pre- and post-amendment comprehensive plan designations. At least 23 TDRs have been certified as of 2020.

Kittitas County, Washington, population 47,935 (2019), adopted a TDR program in 2009 designed to preserve farms, forests and floodplains. Sending site allocations cannot exceed one TDR per 20 acres. Interjurisdictional agreements can regulate transfers to receiving sites within incorporated cities. Receiving sites on unincorporated county land can gain two additional units per TDR in Urban Growth Areas. Outside of Urban Growth Areas, planned unit developments and rural rezones require one TDR per 20 acres. A TDR demonstration project here preserved 480 acres of ranchland.

Lacey, Washington, population 52,592 (2019), uses four city zoning districts as receiving areas for TDRs transferred from Long Term Agricultural sending sites under the jurisdiction of Thurston County. As of 2015, no TDRs had been used in Lacey, reportedly due to little desire to exceed baseline densities.

La Quinta, California, population 41,748 (2019) uses TDR to preserve open space, habitat and other resources in its hillsides. Sending site owners have the option of building on site at a maximum density of one unit per ten acres, if slope and other site conditions allow, or transferring one TDR per ten preserved acres. TDRs can increase baseline density by 20 percent in any residential zone.

Lake County, Florida, population 167,118 (2019), joins point-based regulations with TDR to protect watershed and natural resources near the Wekiva River.

Largo, Florida, population 84,948 (2019) uses TDR to preserve land with archeological, historical, architectural, and environmental significance including wetlands, floodplains and coastal high hazard zones. The city itself has purchased land for open space and banked the development rights for the purpose of granting them free of charge to increase density within affordable housing developments. In the past, developers were not motivated to exceed receiving site baselines. But the program has generated at least one transfer.

Larimer County, Colorado, population 356,899 (2019), lies 50 miles north of Denver, bordering Wyoming. To curb sprawl, Larimer County and the City of Fort Collins adopted the Fossil Creek Reservoir Area Plan in 1998. This plan aims to enhance critical environmental resources, minimize hazards, retain rural character, and maintain community separation between the cities of Fort Collins and Loveland in an area that in 1998 was under county jurisdiction but immediately adjacent to the city limits of Fort Collins.

To prevent development from encroaching into a sensitive natural area, the plan established a ¼ mile natural resource buffer around the edges of the Fossil Creek Reservoir. The plan included a transfer of density units (TDU) mechanism that hinged on intergovernmental cooperation and incorporated motivational features like a cash in lieu option and density incentives for receiving site projects that used the transfer provisions early.

The TDU program was a success, transferring roughly 1,760 TDUs into the 900-acre receiving area. By 2019, the receiving area had been primarily annexed and only one 20-acre property with limited development potential remained undeveloped. Both the city and county acknowledged that the program had reached completion and the TDU components were removed from the Fossil Creek

Reservoir Area Plan.

In this program, the mapped sending area included the Fossil Creek Reservoir and the surrounding land. The standard allocation was 114.5 percent of the density allowed by the underlying zoning. However, this standard allocation could be increased for sending sites with extra community benefits such as environmental resources, wildlife corridors, trail routes, agriculture, park sites, historic landmarks, or scenic views. Similarly, the standard allocation could be decreased if sending sites had existing development, poor location, or development constraints.

The receiving sites were under county jurisdiction but destined for annexation to Fort Collins. Baseline density was one unit per two acres. Developers could use TDUs to exceed that baseline and achieve a density within the range designated in the plan and ultimately determined by the Board of County Commissioners. The TDU program exempted affordable housing units. Developers who chose to use TDU had to use TDU for the entire receiving site. If a site was developed in phases, the TDU could also be acquired in phases.

The TDU requirement was the total number of units proposed, minus the number of units allowed by the baseline density and divided by 1.5. The reduction achieved by the 1.5 division was adopted as a way of motivating developers to use TDU at program start up. This early-adopter incentive was initially supposed to be phased out after two years. However, it was ultimately made a permanent feature of the program when developers argued that transactions would not be feasible without this 1.5 -to-1 transfer ratio.

By intergovernmental agreement, Fort Collins and Larimer County created a program that produced the benefits of an inter-jurisdictional transfer program without actually having to transfer density between jurisdictions. The county and the city jointly planned the receiving areas using city development standards with the understanding that the receiving sites, although in the county, would be annexed to Fort Collins following approval of projects that utilized the TDU option. In fact, the program required receiving site projects to apply for annexation prior to recordation of plats and these properties were annexed as soon as possible following transfer of the TDUs. Even though the receiving site projects were approved by Larimer County, Fort Collins served as a referral agency and conducted its own review of receiving site projects and associated infrastructure improvements.

The program was also an early example of density transfer charges (DTC), allowing receiving site developers to comply by paying cash in lieu of actual TDUs if they met the following three conditions. 1) The receiving site project needed no more than ten TDUs or the receiving site was no larger than 25 acres. 2) A good-faith effort had to be demonstrated that the developer attempted to secure actual TDUs. 3) The developer had to agree to the DTC amount proposed by the county board.

Initially, the formula for calculating the DTC made cash in lieu more expensive than the acquisition of TDUs. That original formula was later changed, making the payment in lieu of one TDU the price estimated for a similar interest in open space times 1.5. Revenues from these in-lieu payments were reserved exclusively for acquiring open space in the sending area.

Lee County, Florida, population 770,577 (2019), adopted its first TDR ordinance in 1986 and has since made several amendments resulting in three separate TDR programs. On Pine Island, which lies in the Gulf of Mexico and offers one bridge connecting to the mainland, transferable development units (TDUs) can be severed from wetland and upland sending sites and transferred at a one-to-one ratio to receiving sites in Pine Island Center or at a two-to-one ratio to residential receiving sites on the mainland as well as to mainland commercial sites at a conversion ratio of 10,000 square feet of floor area per TDU. In a second program for wetlands on the mainland, TDU allocations are doubled for sending sites in Coastal High Hazard Areas. In the Southeast Lee County program, TDUs can be severed from wetlands or density reduction/groundwater resource sending sites. Sites in Coastal High Hazard Areas cannot receive TDUs in order to minimize the number of people who have to evacuate in the event of hurricanes. Under many circumstances, TDU applications can be decided administratively. Lee County saw 500 TDRs severed from sending sites and 200 used at receiving sites between 1986 and 2001. Logically, sending site owners should be motivated by physical constraints as well as regulations, particularly on wetlands, and by the administrative approval process which relieves applicants of the cost, delay and uncertainly of public hearings. However, Lee County developers are often satisfied with the baseline density allowed in receiving zones. In addition, the prohibition on transferring TDUs to Coastal High Hazard Areas has inhibited use of the program since this waterfront land has the highest demand for development.

Lee, New Hampshire, population 4,481 (2018), uses 1:1 transfers between contiguous parcels to preserve farms, forests, watersheds and open space.

Livermore, California, population 89,699 (2019), lies in central Alameda County, 40 miles east of Oakland. In order to preserve its wine country heritage, the city adopted the South Livermore Specific Plan which included a very successful program for mitigating the loss of vineyards plus a TDR mechanism that preserved 213 acres of agricultural land as well as 370 acres of parkland.

The Livermore City Council adopted the North Livermore Urban Growth Boundary Initiative in 2002 aimed at preserving agriculture and open space beyond the Livermore City Limits in Alameda County. The Initiative spelled out special uses and development regulations that would apply to properties that subsequently are annexed to the City. However, the Initiative also included a transferable development credit (TDC) provision that landowners could elect to use regardless of whether they annexed to the City or remained under the jurisdiction of Alameda County.

The TDR ordinance ultimately adopted had eight goals: 1) Retain and expand agriculture; 2) Preserve wildlife, hills, wetlands, natural beauty and peace of open landscape; 3) Prevent further sprawl; 4) Reduce traffic congestion and hazard; 5) Limit air pollution; 6) Avoid extension of facilities and services; 7) Provide for outdoor recreation; and 8) Safeguard Livermore's identity, heritage, and character.

The North Livermore Urban Growth Boundary Initiative established a 14,000-acre TDC sending area and allowed sending area landowners the following TDC allocation options based on their desired level of voluntary participation. One TDC per five acres to landowners who record easements that permanently impose on their properties the use/development restrictions of the Initiative including minimum lot sizes of 40 or 100 acres. Eleven TDCs for each lot voluntarily foregone that would have been allowed by the Initiative. Ten credits for each undeveloped lot deed restricted to remain permanently undeveloped. Twelve credits for each existing dwelling unit removed from a parcel plus permanent restriction of development on that parcel by easement.

The Initiative required the city to create a mechanism to use these TDCs at receiving sites within Livermore. The city essentially created TDC receiving sites wherever a parcel was allowed increased residential density by the 2003 General Plan Update or any future general plan amendments. As also called for in the 2003 General Plan, baseline density in these receiving areas was established as the maximum density allowed prior to the 2003 General Plan Update or any subsequent general plan amendments that resulted in new residential land use designations or an increase in residential density.

Livermore's 2004 TDC Ordinance implemented the North Livermore Urban Growth Boundary

Initiative and the 2003 General Plan Update by establishing three types of receiving areas, all of which establish baseline density as the maximum density previously allowed to the receiving site: 1) TDC combining district; 2) Planned development district; and 3) Zoning districts incorporating TDC. In all three options, property owners who decline to exceed baseline comply with the development requirements applicable to the baseline general plan designation. Owners who choose to exceed baseline must comply with the requirements of the zoning district corresponding to the TDC combining district of the TDC receiving area general plan designation plus all other requirements of the TDC ordinance including the need to acquire TDCs or pay an in lieu fee. In specific plan areas, the specific plan itself was to establish the TDC provisions, if any. Special rules applied for receiving areas with a industrial baseline and a TDC receiving zone designation allowing residential development.

Affordable housing units were exempt from TDC requirements. Market rate units above baseline were required to submit TDCs or in lieu fees as follows.

(1) Two TDCs for each single-family detached dwelling in excess of baseline density (or one TDC for each single-family detached dwelling in excess of baseline density for developments with applications accepted as complete prior to January 26, 2004). (A later code revision changed these requirements as follows: 1.5 TDCs per bonus single-family detached residence at density ranges between one and seven units per acre or 1.25 TDCs for bonus single-family detached residences that secured final map approval by June 1, 2013 and started construction before April 1, 2014; 1.25 TDCs for each bonus single family detached residence in a density range of eight to 14 units per acre.)

(2) One-half TDC for each multi-family attached unit above baseline density.

(3) Payment of the TDC in lieu fee for each required TDC. Based on economic analysis, the in-lieu fee was established as $24,000, meaning $48,000 per bonus single family unit and $12,000 per bonus multiple-family unit. (Revenues from TDC in lieu fees must be used for the acquisition of TDCs from North Livermore. Other than TDC acquisition, revenue from TDC in lieu fees can only be used for costs incurred in administering the TDC program.)

The Livermore TDC Ordinance additionally motivated developers to acquire TDCs in order to proceed with construction within the confines of Livermore's Housing Implementation Program (HIP). Livermore aimed to maintain a desired rate of growth by using HIP to limit the number of dwelling units issued building permits in any given year. However, some HIP allocations could be reserved for developments that further the goals of the general plan including dwelling units in the South Livermore Specific Plan area and the Downtown Specific Plan area, as well as projects that retired TDCs from North Livermore. Specifically, an average of up to 200 TDC-retiring units per year received allocations under HIP from 2005 to 2014. Unused allocations could be carried over to subsequent years, but TDC-retiring units were limited to a total of 2,000 allocations over this 10-year period. Most importantly, TDC-retiring units that received these allocations were not required to compete in the annual HIP process that otherwise would have been required to proceed with a residential project.

Rather than find willing TDC sellers and negotiate sale prices, all developers who used TDC to exceed baseline as of 2010 elected to use the in lieu fee option, resulting in payments totaling $1,576,000. This revenue was combined with other funding sources to purchase an environmentally sensitive property in the North Livermore sending area.

London Grove, Pennsylvania, population 8,752 (2018), uses TDR to preserve farms, woodlands, nature, and places with historic, scenic or cultural resources.

Los Angeles, California, population 3,979,576 (2019), lies at the heart of a six-county region with a population of over 18.5 million. The city has been building a high-density, mixed-use downtown based largely on a plan for the Central Business District (CBD) approved in 1975. The CBD Plan has been implemented in part by three TDR mechanisms designed to achieve a wide range of objectives: preserve historic landmarks, promote affordable housing, create public open space, provide public transportation, and create public/cultural facilities as well as offer flexibility in the concentration of development without overwhelming the overall capacity of the public service and infrastructure system.

As discussed at the end of this profile, as of April 2021, the city was headed toward adoption of a Downtown Community Plan in 2021 that would incorporate a historic preservation TDR component within a community benefits program aimed at prioritizing affordable housing within a multi-level system allowing additional development potential above baseline.

The original CBD Plan allowed unused floor area potential to be transferred within the CBD as long as the donor (sending) and receiving sites were within 1,500 feet of one another and located within four sub-districts. In 1985, the City adopted a variation of the original mechanism called the Designated Building Site ordinance as a way of preserving historic downtown buildings in general and the City's Central Library in particular. When the City adopted a third permutation in 1988, the original 1975 provisions continued to apply to transfers of 50,000 square feet of floor area or less while transfers greater than 50,000 square feet were regulated by the new program, known as Transfer of Floor Area Rights, or TFAR. For many years, TFAR required developers to make a Public Benefit Payment to the City of $35 per square foot of transferred floor area to be used for affordable housing, open space, historic preservation, public transportation and public/cultural facilities. The amount required under the Public Benefits Payment has changed but it is still a highly effective means of generating funding for downtown betterment, as explained below.

Designated Building Sites – In this TDR variation, applicants propose a project encompassing multiple land parcels that are contiguous or separated only by streets or other rights of way. The City Council must find that the Designated Building Site designation is needed to preserve and restore a structure which is designated as historic by the Cultural Heritage Commission as well as City owned and operated. Approval as a Designated Building Site establishes a maximum floor area ratio of 13:1 for the entire land area within the Designated Building Site, not just the receiving site. This development potential must be distributed to preserve the historic landmark while allowing buildings on the receiving site portions of the Designated Building Site to greatly exceed 13:1.

TFAR: Transferable Floor Area Rights – In the TFAR program, sending sites and receiving sites can be any parcels within the CBD. The baseline density allowed to parcels in the project area is either floor area ratio (FAR) 3:1 or 6:1 depending on the subarea in which the site is located. Property owners who do not wish to build up to these baseline densities can sell their unused floor area potential. Conversely, developers who want to exceed these baselines can buy transferable floor area rights, or TFARs, and achieve a maximum FAR 13:1. Developers wanting to exceed baseline can avoid the TFAR requirements only under limited exceptions such as the replacement of an existing building or the development of properties that qualify for a variance due to exceptional circumstances.

Applications for TFAR must meet all of six conditions including plan consistency, appropriateness within the circulation system and compatibility with existing/proposed development as well as the infrastructure system.

Developers are required to pay a Public Benefit Payment on transfers in order to fund public open space, affordable housing, cultural/public facilities, historic preservation and public transportation improvements. Compliance with this requirement can be deferred until the project begins construction. The original flat fee of $35 per transferred square foot of floor area has since been replaced by the following formula: 1) take the sales price or appraised value of the receiving site; 2) divide by the receiving site area; 3) divide again by the site's baseline density limit; 4) multiply by 40 percent; and 5) multiply again by the number of square feet to be transferred to the site. (To illustrate with an actual example, a Public Benefit Payment of $5,273,329 was calculated for the proposed transfer of 242,276 square feet of floor area to a receiving site 34,675.17 square feet in size appraised at $11.3 million within a zoning district with a FAR 6.0 baseline.)

When the donor site is owned by the city, the payment to the city is called the Transfer Payment and is calculated as ten percent of the Public Benefit Payment or $5 per square foot of transferred floor area, whichever amount is greater. In the example from above, the Transfer Payment was $1,213,630. Subject to the city's approval, an applicant can apply a portion of the Public Benefit Payment directly to the actual benefits. For example, the developer in the example above proposed to use $500,000 for pedestrian amenities. The remainder is deposited in a Public Benefit Payment Trust Fund.

Performance

As mentioned above, the Designated Building Site process was created for the Los Angeles Central Library, a beloved landmark built in 1926 and listed in the National Register of Historic Places. In 1985, the Community Redevelopment Agency and a developer jointly applied for Designated Building Site status for five properties in downtown Los Angeles, including the Central Library, with a combined land area of 382,422 square feet. Under the baseline zoning limits, the by-right development on these five parcels would have been limited to a total of 2.5 million square feet of floor area. However, through the Designated Building Site process, a total of more than 3 million square feet of floor area was built including three nearby office buildings, a public plaza and the restoration and expansion of the Central Library itself. It is estimated that, in return for the increased density, the City received an estimated $65-million worth of public benefits from this approval.

As discussed above, the 1988 amendments added a $35 per square foot Public Benefits Payment requirement to larger transfers. Nevertheless, the building boom of the late 1980s resulted in continued use of TFAR. Following the 1988 code changes, three additional projects were approved with a total build-out of 3.5 million square feet of office, hotel and retail floor area including one million square feet of transferred floor area. A study prepared by the New York City Department of City Planning reports that Los Angeles TDR programs have transferred 6.6 million square feet of floor area, resulting in an estimated total of $90 million in community benefits. The New York report also states that the Los Angeles Convention Center has been the largest supplier of TDRs (New York City 2015).

The Public Benefit Payment required by the Los Angeles TFAR process is possibly the most productive transfer surcharge system in the nation. As shown in the example above, one transfer alone can produce millions of dollars in Public Benefit Payment revenue (in addition to proceeds from the Transfer Payment if the sending site is owned by the City). However, it should be remembered that there would be no Public Benefit Payments without a TFAR program that observes key success factors including the following.

- The demand for high rise development here is inevitable given downtown Los Angeles' status as the center of a huge, prosperous metro region. But, unlike many other similarly blessed cities, Los Angeles imposes a baseline density that developers can only exceed through TDR, thereby creating a stream of revenue for public benefits during growth cycles.

- Downtown Los Angeles makes an ideal receiving area given its attraction for employers and extensive infrastructure. It has a tradition of high-rise, mixed use development unlike many other locations in the region where dense development has been resisted by surrounding single-family residential neighborhoods.

- Los Angeles has developed Public Benefits Payment and Transfer Payment formulas that developers appear to be able and willing to pay.

- The city owns sizeable amounts of transferable floor area in the LA Convention Center and other public sites. This serves as an inventory of readily available TFAR, thereby assuring developers that they will be able to buy the floor area they need at a known price. This level of certainty is critical to securing financing for major construction projects which can take five years or more to go from conception to completion.

Draft Downtown Community Plan as of April 2021

Los Angeles is expected to adopt an updated Downtown Community Plan in 2021. In addition to other goals, the draft plan aims to fight climate change by promoting compact centers, preserving/reusing historic buildings, and maximizing green space for the multiple benefits of flood storage, water quality enhancement, habitat improvement, and increased biodiversity. The November 2020 Draft Downtown Community Plan available in April 2021 incorporates a Community Benefits Program allowing developers to exceed baseline development potential up to maximum potential in exchange for community benefits as explained below.

The draft Community Benefits Program for housing developments has three levels of additional development potential above baseline.

- Level One allows up to 35 percent additional FAR and maximum story height in exchange for meeting affordable housing targets.

- Level Two allows additional FAR or stories in exchange for affordable housing, public open space, or community facilities. Within a designated subarea of downtown, Level Two potential can also be gained by transferring TDR, (meaning unused development potential), from preserved historic buildings.

- Level Three allows development potential to achieve the highest possible FAR and story height in exchange for providing additional affordable housing, additional other community benefits (open space, community facilities, TDR from historic buildings) or payment to a Community Benefits Fund.

For non-residential developments, the draft Community Benefits Program offers two levels of additional development potential above baseline.

- Level Two allows non-residential projects to gain additional development potential in exchange for public open space or community facilities. Within a designated subarea of downtown, Level Two potential can also be gained by transferring TDRs from preserved historic buildings.
- Level Three allows non-residential projects to achieve the highest possible FAR and story height in exchange for providing any of the community benefits found in Level Two (open space, community facilities, TDR from historic buildings), or payment to a Community Benefits Fund.

The draft Downtown Community Plan proposes baseline/maximum FARs for nine downtown subareas and baseline/maximum building stories for five downtown subareas. The draft Community Benefits Program explains that the affordable housing benefit could be satisfied on-site, by payment to an affordable housing trust fund, or by participating in affordable housing that is off site but within the Downtown Plan Area. In Level Two, the draft allows one FAR of bonus FAR for each 4 percent of lot area that qualifies as public open space. Also in Level Two, one FAR of bonus FAR is proposed to be allowed for each 2.5 percent of floor area plus 5,000 square feet developed as community facility space. For the historic TDR mechanism, donor sites must be designated at the local, state or federal level or identified in SurveyLA as an individual resource or contributing building to a historic district. Donor historic sites are subject to a preservation contract documenting maintenance, rehabilitation, and inspection requirements plus a covenant acknowledging the reduced floor area.

In addition to the proposed Community Benefits Program, the Fall 2020 draft Downtown Community Plan recommends exploration of a TDR program allowing the transfer of development rights from dedicated public park space in downtown.

Reference
New York City Department of City Planning. 2015. A Survey of Transferrable Developments Mechanisms in New York City. New York: Department of City Planning.

Los Angeles County, California, population 10,039,107 (2019), has had a TDR mechanism operating for decades within its Malibu Coastal Zone, which extends from the City of Los Angeles to the Ventura County border and roughly five miles inland from the shores of the Pacific Ocean. *Our County*, the Los Angeles Countywide Sustainability Plan adopted in 2019, recommends continued use of TDR as a way of steering development away from ecologically sensitive lands and areas most at risk from climate impacts in the future.

Originally, the Los Angeles County TDC program applied to the entire Malibu Coastal Zone. The incorporation of the City of Malibu in 1991 now places about one fifth of the Zone under the jurisdiction of that city and an explanation of that city's TDR program can be found at the profile for Malibu.

The Malibu Coastal Zone is riddled with antiquated subdivisions with small lots approved long before the adoption of modern standards for safety and environmental protection. In 1978, 64 percent of the 13,475 lots of record in the Malibu Coastal Zone were vacant. These lots were and are attractive to those seeking a rural lifestyle within commuting distance of the second-largest city in the US. However, widespread development of these vacant lots would be disastrous on many levels.

The Malibu Coastal Zone is home to a Mediterranean ecosystem supporting remarkable biodiversity. It contains over 900 plant species, over 50 percent of the bird species found in the country, and habitat for several charismatic species such as golden eagles, bobcats, and mountain lions. Partly to safeguard this habitat, much of the Zone gained federal protection by the creation of the Santa Monica Mountains National Recreation Area (SMMNRA) in 1978.

The Zone is famous for natural disasters as well as natural beauty. The area has repeatedly been damaged by drought-fueled wildfires, such as the 2018 Woolsey Fire which killed three people, cost $6 billion, destroyed 1,643 buildings, and scorched 88 percent of the SMMNRA. The steep terrain has limited ability to absorb intense rainfall, particularly after being blackened during the wildfire season, leading to floods, debris flows, and landslides. Between 1992 and 1995, Malibu was declared a federal disaster area five times due to wildfires, floods and landslides.

The topography here resists the creation or expansion of urban infrastructure. This has made the Zone a difficult place for the effective deployment of fire/rescue vehicles. Existing narrow, twisting roadways can easily be blocked just as residents are trying to flee and emergency vehicles are attempting to access areas threatened by fires, floods, and mudflows.

The lots in the sending areas typically range between 4,000 and 7,000 square feet each and rely on septic systems. When septic systems fail, aquatic environments are damaged. In the 1970s, coliform standards were exceeded in two streams that enter the Pacific Ocean at some of the most popular beaches in Southern California.

The California Coastal Zone Conservation Act, established by voter initiative in 1972, and the subsequent California Coastal Act of 1976 aim to protect coastal resources and promote public access to the ocean. The California Coastal Commission established land use regulations for the Malibu Coastal Zone. Under the Act, new subdivisions could only be permitted if more than half of the lots in existing subdivisions were developed. Well over half of the lots in the zone were vacant in 1978. But strict observance of the subdivision policy would prevent new subdivisions that are properly located and meet modern environmental standards while allowing the development of thousands of substandard, vacant lots that the Commission wants to eliminate in order to reduce environmental damage and limit the number of people vulnerable to wildfires and other disasters.

As a solution, the Coastal Commission adopted a policy of requiring one TDC for each new lot or dwelling unit allowed in the Malibu coastal plain, the receiving area. In 1981, the allocation of TDCs in the sending areas expanded to include resource areas and well as substandard lots. But the overarching goals remain: to reduce development potential in antiquated subdivisions, environmentally sensitive land, and hazard-prone areas without increasing the total development capacity of the Malibu Coastal Zone.

As of a comprehensive reworking of the county's zoning code in 2019, Section 22.44.1230 reaffirms receiving area identification by requiring one TDC, (representing the retirement of one or more sending area lots), for each new lot or each new dwelling unit in a multi-family development, senior complex, or habitable accessory structure.

Sending areas include two categories of rural villages. The primary areas are lots in eight antiquated subdivisions and there are three-plus ways of determining TDC allocations. 1) One TDC can be allocated to a legal lot with a total credit area of at least 1,500 square feet that is served by an existing road and water main and not located in an area of landslide or other geologic hazard. The total credit area is calculated using a formula involving average slope, contour intervals, total length of contour lines, and area of the building site. As an alternative, the required 1,500 square feet of credit area can be assembled by adding 500 square feet of credit area per lot from lots that are at least 4,000 square feet in area, not subject to geologic hazards, and within 300 feet of an existing road or water main. 2) One TDC can be granted for the retirement of any combination of legal lots totaling at least one acre regardless of road and water service. 3) Another two calculation variations are available in the Monte Nido subdivision.

The secondary sending sites in the rural village category are located in five antiquated subdivisions where TDCs are granted for the retirement of at least two contiguous lots containing H-1 and/or H2 habitat. Alternatively, TDC allocation can be determined by using the calculations available to the primary rural villages.

In addition to rural villages, sending sites can be approved for parcels up to and including 20 acres if at least half of the land area is H2 habitat, or the parcel is within 200 feet of H1 habitat or public parkland. Parcels meeting these criteria that are larger than 20 acres qualify for one TDC per 20 acres and/or fractional TDCs.

To be granted TDCs, sending site owners must record an open space deed restriction and either convey title to a public entity or merge the retired lots with an adjacent buildable lot using code provisions for lot mergers, lot line adjustments, or reversion to acreage.

The expansion of the program to include resource lands as well as antiquated subdivisions occurred in 1981. In addition, the California Coastal Conservancy committed to the success of this program by forming a TDC bank using $2.6 million to buy 213 TDCs from sending sites in four antiquated subdivisions.

The El Nido subdivision was created in the 1920s with narrow winding roads and 347 lots on 70 acres of land. By 1980, only 40 lots were developed. Los Angeles County had inherited 153 lots due to property tax default and it was offering them for sale to the public whether or not they were buildable. Approximately 25 of the lots were in or near the bed of a creek that drains into a canyon which is now a public park. Through the Conservancy, the 153 lots owned by Los Angeles County and 30 other lots were permanently retired.

The Conservancy's second restoration project was the Malibu Lake small lot subdivision, which is now surrounded on three sides by Malibu Creek State Park. This tract was created in the 1920s and 1930s to provide cabin sites adjacent to a private hunting camp. Within the Coastal Zone, only 16 of the 158 lots were developed in 1981. Many of the lots were not suited to septic systems and further development of these lots could threaten the quality of the water in Malibu Creek. The Conservancy purchased 125 lots here for $773,000.

The third Conservancy project was the Cold Creek Watershed, a 5,000-acre area containing exceptional wildlife habitat supported by one of the few perennial streams in the Santa Monica Mountains. The Coastal Commission originally required that TDCs used on receiving sites within the Cold Creek watershed had to come from sending sites within the Cold Creek watershed. The only small-lot subdivision within the Cold Creek project area is the Monte Nido subdivision, a 1926 tract with 416 lots on 40 acres. Although the lots average only 4,000 square feet, each lot relies on individual septic systems, including the lots immediately adjacent to the two blue-line streams that cross the subdivision.

The fourth Coastal Conservancy project is the Las Flores Heights Restoration Program. In 1918, the Las Flores Heights subdivision was created with 102 lots on 160 acres. The lots in Las Flores Heights range from one-half acre to an acre in size, but they tend to be steep, and many are not served by a paved road. This area is particularly susceptible to natural disaster. In the 1930s, 20 homes existed in this subdivision; however, by 1982, all but six of the homes had been eliminated by fires and floods. The fires of 1993 reduced that number even more. The Conservancy granted the Trust $886,000 to acquire a major interest in a landholding which included 60 percent of the Las Flores Heights subdivision plus a 160-acre site to the north of the tract.

The Monte Nido subdivision was considered by some to have too few potential sending sites to create an adequate supply of reasonably-priced TDCs. The TDCs from Monte Nido were priced much higher than the TDCs in other parts of the Malibu Coastal Zone. In addition, there were concerns that the owners of the potential sending sites could cooperate to drive up TDC prices even more or perhaps block a proposed development. One developer stated that this requirement was essentially a denial of his project, claiming that it would be difficult or impossible to buy enough TDCs from within the watershed to mitigate a proposed subdivision. In response, the Coastal Commission allowed a tract outside of the Cold Creek watershed, the Fernwood small lot subdivision, to serve as a reserve source of TDCs for receiving sites within the Cold Creek watershed. Fernwood is the largest of the small lot subdivisions, with 1,497 lots, 1,154 of which were undeveloped in 1979.

To further facilitate transfers, developers were allowed to make payments in lieu of actual TDCs which the Conservancy used to buy TDCs. The Coastal Commission also capitalized the Mountains Restoration Trust, a non-profit satellite organization of the Conservancy that could purchase TDCs at below market rates using creative techniques that are not always available to a governmental agency.

The Conservancy started the Mountains Restoration Trust with a $300,000 grant for the purchase of TDCs. Five percent of the in-lieu fee was to be reimbursed to the Conservancy until the grant was fully repaid. However, the demand for Cold Creek TDCs declined. Instead of buying TDCs, the Trust found itself accepting TDCs as donations from homeowners wanting charitable-donation tax benefits in exchange for scenic easements. Typically, these donations were made by homeowners who owned five contiguous lots but only used two or three of these lots as a building site. By donating scenic easements on the two or three undeveloped lots, these property owners were able to continue to use these lots as private open space, yet they received tax benefits as high as $150,000. Using this process, the Trust accepted easements from over 46 lots, representing 24 TDCs. Because these TDCs were acquired for little or no money, the Trust was able to sell TDCs for $15,000 to $18,000, a fraction of the price originally assumed.

In "Transfer of Development in the Malibu Coastal Zone", Elizabeth Wiechec explains that the program was modified in 1984 to reduce the ability to grant TDCs for the retirement of lots which were actually unbuildable due to geologic hazards, septic system limitations and flood hazards. In another modification, the Coastal Commission refined its maps of Environmentally Sensitive Habitat Areas (ESHAs) to protect only the riparian area flanking streams. As a result, many properties which were previously considered unbuildable became viable home sites with commensurate increases in value. This greatly decreased the likelihood that they would become TDC sending sites (Wiechec).

The program languished until 1987 when the coastal zone plan was adopted and formally memorialized the TDR mechanism. For the next five years, the program was extremely active. The work of the Coastal Commission managed to keep the cost of the average TDC at about $20,000, or roughly two percent of the price of new lots in the receiving area. In addition, the subdivision of 10-acre parcels into 2.5-acre lots in the highly desirable receiving area was capable of three- and four-fold increases of per acre value.

By the time the City of Malibu incorporated in 1991, the TDR program had protected an estimated 800 acres of sending area land in antiquated subdivisions and retired roughly 924 vacant lots. By 2003, the Mountains Restoration Trust had retired the development rights on 260 acres of land within the Cold Creek watershed, representing 22 TDCs. In addition, the Trust had collected in lieu fees equivalent to 39 additional TDCs, resulting in a grand total of 544 TDCs at that time (Wiechec).

When Malibu's incorporation became effective in 1991, the city imposed a building moratorium and demand for TDCs fell. The California Coastal Commission prepared and adopted a Local Coastal Plan (LCP) for Malibu in 2002 with a TDR program based on the program that existed when Malibu was part of unincorporated Los Angeles County. The City sued the state over this action but lost in court. In 2004, the City adopted its own LCP which included a TDR program that closely adheres to the Commission's 2002 program.

The City of Malibu's TDR program is profiled separately. However, it deserves mention that the success of the Los Angeles County program is highly dependent on Malibu because the most suitable receiving sites are within Malibu. As of 2010, most of the larger vacant parcels in Malibu were zoned for 20- or 40-acre minimum lot sizes, leaving little opportunity for subdivision. Although some five-acre parcels are zoned for a one-unit-per-acre density, environmental constraints create significant hurdles to subdivision (Edmondson 2010). As mentioned above, the 2019 Los Angeles County Sustainability Plan proposes TDR as a way of steering development away from ecologically sensitive lands and areas most at risk from climate impacts in the future. The implementation of that 2019 plan is just beginning as of 2021.

References

Edmondson, Stefanie. 2010. E-mail correspondence with the author of July 13, 2010.

Wiechec, Elizabeth. Transfer of Development in the Malibu Coastal Zone. Unpublished paper prepared for Mountains Restoration Trust by former Executive Director.
http://www.amlegal.com/nxt/gateway.dll?f=templates&fn=default.htm&vid=amlegal:lapz_ca.

Los Ranchos de Albuquerque, New Mexico, population 6,108 (2019), uses TDR to preserve agriculture and village character. One of the receiving areas is the Village Center, where TDRs transferred at a one-to-one ratio allow developments to exceed a baseline of 12 units per acre by ten additional units.

Lower Chanceford Township, York County, Pennsylvania, population 3,112 (2018), uses TDR to preserve high quality land within its agricultural and conservation zones.

Lumberton Township, Burlington County, New Jersey, population 12,205 (2018), is located 25 miles east of Philadelphia and 40 miles west of the shores of the Atlantic Ocean. The township's first TDR program was a success, largely because of an administrative approval process made possible by comprehensive requirements that assured compatible receiving site projects without the delay, modifications and costs often generated by procedures involving discretionary decisions.

The State of New Jersey adopted the Burlington County TDR Demonstration Act in 1989 in order to test the feasibility of this preservation tool. Lumberton Township became the first jurisdiction to use this act by adopting a program in 1995 known as TDR I. The program aimed to conserve agriculture and other rural resources. The mapped 1,513-acre sending area was land designated as Rural Agriculture TDR Sending Area (RA/S) in western Lumberton Township. Sending sites had to be six or more acres in size and assessed as farmland.

The TDR allocation was based on the maximum density allowed by zoning, which was one unit per two acres, plus any reductions caused by environmental constraints, which essentially meant suitability for on-site septic systems. The exact formula was the number of acres of slightly-constrained soils divided by 2, plus the number of acres of moderately-constrained soils divided by four, plus the number of acres of severely-constrained soils divided by six, multiplied by 1.1 equals the number of TDRs available to the sending site. Fractional results were rounded down and one TDR was subtracted per existing dwelling unit.

Property owners could appeal their credit allocation using either a new soils survey or by submitting a subdivision plan with soil borings indicating that the site had more development potential than the formula-based allocation. However, TDR credit allocations could not be contested after the owner had recorded a conservation easement on the property.

The TDR easement restricted further development of the sending site to one unit per 50 acres. The Lumberton code had an elaborate procedure allowing credits to be reattached to a sending site in the event that the owner was unable to sell the TDRs or for "good and sufficient reasons the public interest would be served by allowing relief from the restrictions imposed by the TDR program."

The receiving area included 508 acres of land mapped as receiving areas in the township master plan. In 1999, the receiving areas and density bonus allowances were changed to the following. 1) One TDR allowed one additional single-family detached unit in the RA/R-1, RA/R-3, and RA/R-5 zones. 2) One age-restricted, attached unit in the RA/R-4 zone was allowed per 0.6 credit. 3) One age-restricted, attached unit in the RA/R-6 zone was allowed per 0.7 credit. The maximum density allowed with TDR ranged from one unit per acre to five units per acre depending on the zoning of the receiving site.

Significantly, the use of TDR was granted administratively as long as the receiving site project adhered to all development requirements, including 35 pages of standards and guidelines found within the TDR code section itself addressing the retention of natural elements and cultural features, as well as the provision of storm water management facilities, public utilities, and landscaping, Since the receiving area was adjacent to the Historic District of Lumberton Village, the TDR code also required new buildings to be compatible in scale with existing structures and reflect the architectural styles of the 18th, 19th and early 20th centuries including the Colonial, Georgian, Federal, Greek Revival, and Victorian styles. To emphasize this requirement, the TDR code incorporated 17 pages of architectural standards regarding façade treatment, building materials, fenestration, rooflines, fences, and open space. While this level of detail is unusual in a TDR code, it may also explain why Lumberton was able to approve receiving site projects without the discretionary review that can cause the delay, redesign, and added costs that sometimes discourage developers from choosing the TDR option.

In 2000, Lumberton adopted a TDR II program that added 1,355 acres of sending area and 185 acres of receiving area. The receiving area zoning for TDR II, RA/R6, allowed one extra age-restricted dwelling unit per TDR. Property owners and developers declined to participate in the new program and TDR II languished. In 2018, all TDR provisions were removed from the code. However, by that time, TDR had succeeded in preserving 850 acres of farmland in Lumberton.

Lysander (Town), Onondaga County, New York, population 22,877 (2016), adopted TDR to preserve farmland. The TDR code was subsequently replaced by Incentive Zoning, Article XXVII, which allows changes in density, floor area, lot coverage, setback, height, buffers, or land use to developments that offer certain public benefits including preserved farmland, parks, open space, utilities, cultural/historic facilities, or a cash-in-lieu payment to acquire these benefits.

Madison, Georgia, population 4,210 (2019), adopted a TDR program to save natural areas, habitat, greenspace, trails, gateways, parks, and historic landmarks. The ordinance specifically allows city owned parcels to qualify as sending sites. Madison severed TDRs from existing parkland and other public open space, transferred 32 of the resulting TDRs to an age-restricted receiving site project and deposited the remaining four TDRs in the city's TDR bank.

Malibu, California, population 12,620 (2019), is an affluent coastal city located 15 miles west of downtown Los Angeles. Before its incorporation in 1991, what is now the City of Malibu served as the primary receiving area for a TDR program established by the California Coastal Commission in 1981 to conserve significant environmental areas, reduce human exposure to wildfires, floods, and landslides, and limit the development potential possible on thousands of substandard lots within antiquated subdivisions located in the steep terrain of the Santa Monica Mountains. For details on the precursor to Malibu's TDC program, see the profile on Los Angeles County.

By the time the City of Malibu incorporated in 1991, the Los Angeles County TDR program had protected an estimated 800 acres of sending area land in antiquated subdivisions and retired roughly 924 vacant lots. By 2003, the Mountains Restoration Trust had retired the development rights on 260 acres of land within the Cold Creek watershed, representing 22 TDCs. In addition, the Trust had collected in lieu fees equivalent to 39 additional TDCs, resulting in a grand total of 544 TDCs at that time (Wiechec).

When Malibu's incorporation became effective in 1991, the city imposed a building moratorium and demand for TDCs fell. The California Coastal Commission prepared and adopted a Local Coastal Plan (LCP) for Malibu in 2002 with a TDR program based on the program that existed when Malibu was part of unincorporated Los Angeles County. The City sued the state over this action but lost in court. In 2004, the City adopted its own LCP which included a TDR program that closely adheres to the Coastal Commission's program.

In Chapter 7 of the City of Malibu's LCP Local Implementation Plan, the goals of the TDC program stress the aim of reducing risks to life and property due to high geologic, flood and fire hazards and the increased amount of erosion that would result from roads, utilities, and building pads needed for new development as well as mitigating the impact on environmentally sensitive habitat, scenic and visual resources, natural landforms, and potential recreational uses.

The Malibu Local Implementation Plan identifies six types of sending areas: 1) existing lots within seven named small lot subdivisions; 2) existing lots within four other small lot subdivisions that have environmentally-sensitive habitat and are contiguous to each other or to other retired lots; 3) parcels consisting primarily of environmentally-sensitive habitat; 4) parcels located within eight significant watersheds; 5) parcels adjacent to existing parkland in the Santa Monica Mountains where development cannot be located due to encroachment of fire abatement requirements onto parks; and 6) parcels in wildlife corridors in the Santa Monica Mountains Coastal Zone.

Within each of these sending areas, the TDC allocation formula requires: 1) all land included in the calculation of credit area must be in legal lots; 2) the number of credits allocated is the square footage of the sending lot times the product of 50 minus the average slope of the lot; 3) in the four small lot subdivisions, at least three lots must be retired, the lots must be contiguous to each other or other retired lots, and all the lots must contain significant habitat.

These credits can be transferred to any part of the City of Malibu where new lots can be created via subdivision or where multi-family residential units are permitted. In single-family subdivisions and larger multiple-family residential projects, one TDC is needed for each new lot or additional dwelling unit created. Affordable housing units are exempted.

As of 2009, Malibu reported that two TDCs had been transferred allowing two existing parcels to be split into four lots. Three transfers were pending as of December 2009 requiring eight more TDCs. One reason given for the low level of activity was attributed to Malibu's goal of remaining a community with low density development (Walls & McConnell 2007). In a more nuanced explanation, a Principal Planner with Malibu reported in 2010 that the city does not receive many subdivision applications because there are few parcels with zoning designations that would allow subdivision. Malibu's larger parcels are zoned for minimum lot sizes of 20 or 40 acres due to rezonings adopted by the city when it incorporated in 1991 and by the Coastal Commission when it adopted the Local Coastal Program in 2002. There are some five-acre parcels zoned for one-acre minimum lot size but the findings for a subdivision are difficult due to prohibitions against the disruption of Environmentally Sensitive Habitat Areas and the subdivision of any beachfront property that would require a seawall, bulkhead, or some other form of shoreline protection. Because there is little subdivision potential, the demand for TDCs is low (Edmondson 2010).

References
Edmondson, Stefanie. 2010. E-mail correspondence with the author of July 13, 2010.

Wall, Margaret, and Virginia McConnell. 2007. Transfer of Development Rights in US Communities. Washington, D.C.: Resources for the Future.

Wiechec, Elizabeth. Transfer of Development in the Malibu Coastal Zone. Unpublished paper prepared for Mountains Restoration Trust by former Executive Director.

Manheim Township, Lancaster County, Pennsylvania, population 40,232 (2018), lies within Lancaster County, 75 miles west of Philadelphia. To limit development on prime agricultural land, the Township adopted an agricultural land preservation program in 1990. Under this program, major portions of the Township were rezoned from a residential zone allowing almost three dwelling units per acre to an Agriculture Zoning District which allows one unit per 20 acres. As compensation for that reduction in development potential, the Township, in 1991, introduced a transfer of development rights program.

Process

The TDR section of the Manheim Township Zoning Ordinance is designed to preserve prime agricultural soils and the agricultural character of the land by shifting development from sending areas to receiving areas. The Agricultural District is established as the sending area and the R-1, R-2 and R-3 Residential Districts plus the T-Zone Overlays are established as the receiving areas.

Manheim Township allows severed TDRs to be transferred immediately to a receiving site or held by a purchaser for future use or sale. The price of the development rights is determined by willing buyers and sellers. The Township itself can purchase development rights and accept rights as gifts; these development rights may be retired or sold by the Township. When development rights have been severed, a deed restriction is placed on the sending site allowing only agricultural uses. This deed restriction, called a Declaration of Restriction, designates the Township as a third-party beneficiary.

Sending sites are parcels in the Agricultural District of at least ten acres in size. The number of TDRs available for transfer is calculated by multiplying the total number of acres of unencumbered land subject to the Declaration of Restriction times 0.73.

Each TDR yields one dwelling unit above baseline density in qualified residential receiving sites. By using TDR, maximum density can go from 2.2 to 2.9 units per acre in the R-1, from 2.9 to 4.3 units per acre in the R-2 (or 7.0 units in a PRD), and from 5.0 units per acre to 6-10 units per acre in the R-3 zone for PRDs and specified development within the T-5 Oregon Village Overlay Master Site Plan.

In addition to density, developers can use TDR to exceed baseline building heights in the R-3 zone. Specifically, one TDR is required for each dwelling unit above 35 feet in height when using conventional development provisions and one TDR is required for each dwelling unit above 40 feet in height developed under PRD provisions.

On qualified nonresidential or mixed-use receiving sites, one TDR allows 3,000 square feet of nonresidential floor area or one dwelling unit above the baseline height. In five zones, one TDR can permit 5,000 square feet of floor area beyond baseline building length.

In the master site planned development (Oregon Village Overlay), one TDR is needed for each three acres of master site planned development.

Program Status

The Manheim Township program has the ingredients to motivate transfers. Owners of sending sites can only achieve a density of one unit per 20 acres on site; conversely, those owners can transfer development rights to receiving sites at a density of 0.73 units per acre, representing a substantial transfer ratio of 14.6:1. Receiving site owners can use TDRs to increase residential density or, in some zones, exceed building height and length baselines.

In the early years of the program, Jeffrey Butler, Manheim Township's Director of Planning reported that 124 development rights were severed, preserving approximately 170 acres. Most of these severed rights were purchased by the Township itself using general fund money and contributions from developers. The Township used the proceeds of selling TDRs to developers at auction as a revolving fund to buy and bank additional TDRs.

Lancaster Farmland Trust has awarded Manheim Township with its Amos Funk Spirit of Cooperation Award for preserving farmland through TDR. The Trust and Manheim Township have cooperated in the preservation of four farms with a combined acreage of 300 acres. At the time of that award, the Trust acquired ten percent of a farm's TDRs and Manheim Township itself purchased 90 percent. Both the Trust and the Township resold these TDRs and used the proceeds to make additional purchases.

In its study of TDR programs in Pennsylvania, the Brandywine Conservancy notes that Manheim Township astutely minimized the potential for NIMBYism in its TDR program by designating Kissell Hill as a receiving area, which is bordered by various non-residential uses.

A 2015 paper by Preston Hull reported that Manheim initially jumpstarted the program by buying and banking 242 TDRs. Developers found it easier to buy TDRs from the bank rather than deal directly with sending site owners; as a result, the private market originally intended by Manheim was slow to materialize. In total, Manheim purchased 377 TDRs for between $4,500 and $6,500 each before suspending bank purchases in 2006. According to Hull, when developers bought TDRs directly from farmland owners in 2015, the prices averaged from $10,000 to $12,000 per TDR and Manheim experienced 251 direct transfers between developers and farmers as of April 2015.

By October 2020, 871 TDRs had been severed, preserving almost 1,200 acres of farmland.

Mapleton, Utah, population 10,168 (2018), uses TDR to preserve agricultural land, rural open space, scenic vistas, sensitive lands, natural hazard areas and places where delivery of public services would be difficult and/or expensive, such as hillsides and mountainsides. The transfer ratio is one-to-one except in the Critical Environment (CE-1) zone applicable to hillsides, where transferable TDRs are three times baseline density for sending sites that remain privately owned upon the recording of a conservation easement or five times baseline when ownership is conveyed to the city. These bonuses in the CE-1 can only be used by sending parcels that were not developed or subdivided after the 1998 adoption of the TDR program. Receiving areas include land within seven different zones. The city may approve the use of TDRs in conjunction with upzoning requests that are consistent with the comprehensive plan. Mapleton issued 399 TDRs to sending sites between 1998 and 2010 when the city stopped accepting sending site applications. These TDRs preserved approximately 750 acres. Between 1999 and 2019, the developers of 32 subdivisions used 209 TDRs to increase receiving site density by roughly 400 dwelling units.

Marathon, Florida, population 8.581 (2019), encourages the preservation of sending sites with Class 1 habitat by allowing the transfer of development potential to receiving sites not classified as Class 1 habitat.

Marin County, California, population 258,826 (2019), uses TDR to protect agricultural land and environmentally sensitive areas in places where it has been approved as a planning implementation measure. One source reports that TDR had preserved 670 acres here as of 2007.

Marion County, Florida, population 365,579 (2019), motivates owners of agricultural and environmental sites of at least 30 acres to transfer one TDR per preserved acre. Three types of mapped receiving areas surround the City of Ocala. At least 3,580 acres are preserved. In 2013, the county adopted another program to transfer credits from antiquated subdivisions without central water.

Mequon, Wisconsin, population 24,382 (2019), adopted a TDR ordinance aimed at conserving agriculture, forestry, open space, natural hazard zones and sensitive environmental areas. The program placed a conservation easement on a 112-acre parcel of land abutting the Mequon Nature Preserve.

Miami, Florida, population 467,963 (2019), promotes transferable development potential under a Public Benefits Program aimed at generating workforce housing, public parks, and historic preservation. Receiving site projects can achieve additional building height and floor area by providing actual improvements or making a cash in lieu payment to the city's Affordable Housing Trust Fund and the Parks and Open Space Trust Fund.

Miami-Dade County, Florida, population 2,716,940 (2019), stretches from the dense coastal cities of Miami and its suburbs into the wetlands and mangrove forests of the Florida Everglades, which occupy the western and southern half of the county. As development threatened irreparable damage, Everglades National Park was established in 1934 and is now the third largest national park in the contiguous United States. The park is home to hundreds of bird, fish, reptile, and mammal species, including dozens of threatened or endangered species like the Florida panther and West Indian manatee, which largely explains the park's listing as a Biosphere Reserve, World Heritage Site, and Ramsar Wetland of International Importance.

In addition to preserving biodiversity, the Florida Everglades protects the recharge area for the aquifer supplying drinking and irrigation water to most of southern Florida and safeguards freshwater flow to the largest mangrove ecosystem in the Western Hemisphere. Over long time periods, mangroves and coastal wetlands in general are able to store five times more carbon than tropical forests. Conversely, coastal wetlands release massive amounts of long-sequestered carbon when they are degraded or destroyed. Drawdown calls for the protection of the Florida Everglades in particular and wetlands in general for protecting wildlife, water quality, and storm resilience as well as sequestering carbon (Hawken 2017).

In 1981, Miami-Dade County adopted its East Everglades Ordinance, which designated a 242-square mile area contiguous with Everglades National Park area as an Area of Critical Environmental Concern in order to protect water quality, flood storage capacity, biodiversity, and the economic vitality of the county and its municipalities. This ordinance included TDR provisions, which in Miami-Dade County are called severable use rights, or SURs.

The sending area for this program included Management Areas 1, 3B and 3C of the East Everglades area, with a combined area of about 45,200 acres. SUR allocations differ between these three sending areas. Management Area 1 is partly developed, and property owners here can build on site at a maximum density of one unit per 40 acres although with stringent development requirements, including provisions to prevent disruption in the natural flow of surface water. However, when owners of property in Management Area 1 protect their land by permanent easement, they can sever and sell the foregone development potential at the ratio of one SUR per five acres. The allocation ratio is one SUR per 12 acres in Management Area 3B and one SUR per 40 acres in Management Area 3C. Owners of legal lots smaller than these minimums can sell one SUR per lot if they registered within one year of program adoption. Lands traditionally submerged at least three months per year are considered undevelopable and cannot sell SURs.

Receiving areas are properties with residential, commercial, and industrial zoning designations located within Miami-Dade County's Urban Development Boundary (UDB). All Miami-Dade County zones designated for urban development can receive SURs with the exception of the agricultural, environmental, recreation, and open space zones. In four residential zoning districts, developers can use SURs to reduce minimum lot size and minimum frontage. In four additional residential zoning districts, SURs can be used to reduce maximum coverage as well as minimum lot size and frontage. In three more residential zones, SURs can increase maximum height limits as well as density, floor area ratio, and coverage. In two additional residential districts, SURs can increase maximum height, density and floor area ratio but not coverage. In the PAD, ECPAD, and REDPAD districts, SURs can increase density by 20 percent. In the Core or Center Sub Districts of Community Urban Center zoning districts with certain designations, baseline density can be increased by up to eight units at the rate of two units per SUR. In seven commercial districts, baseline FAR can be exceeded at the ratio of 0.015 FAR per SUR in seven commercial districts and by 0.010 FAR in the OPD district.

A 2017 report explains that some zones introduced after the SUR program launched have smaller minimum lot sizes than receiving site projects can achieve with SURs. For example, the RU-IZ, RU-1M(a) and RU-1M(b) zones had minimum lot sizes of 4,500, 5,000 and 6,000 square feet respectively, while the smallest lot achievable by SUR was 6,000 square feet. As a result, developers could circumvent the use of SURs by applying for a rezoning to any of these three zones (Miami-Dade County 2017).

The 2017 report also found that SURs had not been used in Core and Center Subdistricts of Urban Centers because the code allowed considerable density there by right. The report noted that demand for SURs could be generated by reducing the baseline densities in these zones (Miami-Dade County 2017).

County policy required the use of SUR or PDR for any expansion of the UDB involving residential but not non-residential development. The 2017 report observed that the SUR program could be strengthened if the policy both specified the extent of SUR/PDR required for UDB expansion and made this requirement apply to all types of development rather than just residential development (Miami-Dade County 2017).

The receiving areas are subject to annexation by incorporated cities, thereby reducing the demand potential within the county. In response, as of 2015, Dade County allows SURs to be transferred inter-jurisdictionally to incorporated cities that agree to participate. The 2017 report suggests that it may be necessary to require or encourage incorporated cities to create receiving areas for the interjurisdictional transfer of SURs (Miami-Dade County 2017).

Importantly, receiving site developers can use SURs at receiving sites as a matter of right, which can reduce developer concerns about the delays, changes, and cost increases that sometimes result from discretionary approval.

A sizeable supply of SURs has been created here because the U.S. Army Corps of Engineers and the Department of the Interior do not include the value of SURs when these agencies buy property in the East Everglades. Consequently, these former property owners have SURs to sell. At one time, receiving area developers were able to buy SURs directly from these former property owners for roughly $2,500 each or from intermediaries for between $3,000 and $5,000 each. The 2017 report finds that this practice dilutes the value of SURs and consequently makes the TDR option less attractive to the private owners of sending area property. The report notes that the problem of low SUR prices could be addressed if the federal agencies acquired and extinguished SURs when they purchase private property, actions that would reduce SUR supply and likely increase the value of the remaining SURs (Miami-Dade County 2017).

As of January 2016, 1,116 of the 4,700 SURs available to sending sites had been applied to receiving sites. Most of the transfers have gone to residential developments. An average of 54 SURs per year were applied between 1989 and 2015, with peak years occurring in 2000 and 2006. SUR use dropped during the Great Recession and started to pick up again in 2014. A 2016 staff report recommended that the county explore the possibility of supplementing the SUR program with an agricultural TDR program and a historic resources TDR program.

References
Hawken, Paul. 2017. Drawdown: The Most Comprehensive Plan Ever Proposed to Reverse Climate Change. New York: Penguin Books.

Miami-Dade County. 2017. Report Evaluating Existing and Potential Development Density Transfer Programs -Directive No. 152550. Memo to Board of County Commissioners dated January 23, 2017.

Middle Smithfield Township, Monroe County, Pennsylvania, population 15,647 (2016), promotes the preservation of farmland, historic resources, and environmentally sensitive areas by allowing residential receiving sites extra density and commercial projects additional impervious lot coverage in return for preserving properties in two zoning districts.

Milton, Georgia, population 38,759 (2019), uses the TDR mechanism embodied in SmartCode to preserve environmental areas, rural character, parks, greenways and trails throughout the city including mapped greenspace within two districts regulated by form-based codes: Crabapple and Deerfield. In Crabapple, each transferred TDR allows two bonus building units and each building unit represents one dwelling unit or 2,250 square feet of non-residential floor area.

Minneapolis, Minnesota, population 429,606 (2019), uses TDR to transfer unused floor area from downtown historic structures that are rehabilitated and subjected to the restrictions that apply to designated landmarks. Receiving sites can use transferred floor area to increase baseline floor area limits by 30 percent. The Planning Director's administrative decisions are final unless appealed. The Handicraft Guild Building was the first landmark to be protected by this program.

Monroe County, Florida, population 74,228 (2019) uses TDR to shift development away from environmentally sensitive sites that are also vulnerable to coastal storms since the entire county is within the 100-year flood plain. Owners of potential sending sites are motivated to use the TDR option by the difficulty and expense of attempting to build on site due to physical and regulatory constraints. However, receiving sites are also limited in their ability to accommodate transferred density. As of 2014, 99 credits landed in 32 projects.

Monterey County, California, population 434,061 (2019) allows development potential to be transferred from parcels impacted by restrictions in the "critical viewshed" that protects views of the Big Sur coastline from scenic Highway 1.

Montgomery County, Maryland, population 1,050,688 (2019), extends north from the District of Columbia and includes many of the suburbs that have absorbed the growth generated by adjacent Washington, D.C., including Silver Spring, Bethesda, and Rockville. As of 2013, Montgomery County had permanently preserved 72,479 acres of farmland, with over 52,052 acres of this total being accomplished by TDR alone, representing a value of $117 million.

In 1964, the county adopted a general plan entitled *On Wedges and Corridors*. The plan aimed at concentrating growth in a central spine served by public transportation and urban infrastructure while retaining the remaining wedges as farmland and open space. Initially, the county tried to implement this plan using only zoning. However, after a one-unit-per-five-acre downzoning was adopted in 1970, almost 12,000 acres of agricultural land was lost in the next five years.

The county concluded that farmers were reacting to a condition called "impermanence syndrome", a feeling of resignation about the inevitability that the county would eventually become a homogenous suburb. According to this syndrome, if neighboring land is sold for development, adjacent land is rendered less viable or even impossible to farm due to complaints from subdivision residents about odors, dust, chemicals, traffic delays, and other land use conflicts common to the farmland/subdivision interface.

The 1980 *Functional Master Plan for the Preservation of Agriculture and Rural Open Space* and its subsequent zoning code amendments established a 93,000-acre Agricultural Reserve and reduced the maximum density allowed on site to one unit per 25 acres, the smallest amount of land considered by the county to be capable of supporting a commercial farm. Owners of land in the Agricultural Reserve could decline to use the TDR option. But if they chose to participate, they would record a conservation easement permanently binding the density limit of one unit per 25 acres. For every five acres of sending area land placed under easement, participating property owners were issued one TDR.

Unlike many TDR programs that distinguish between farmland and natural lands, or in some cases between properties with different agricultural productivity or development capability, Montgomery County reasoned that the integrity of the entire landscape, natural as well as cultivated, deserved preservation, which is why the entire 93,000-acre Agricultural Reserve was eligible to participate. This philosophy simplified the program and made it appealing to owners of various kinds of land. The uniformity of eligibility throughout the Ag Reserve also addresses the concern about impermanence syndrome since it prevents interspersed development from introducing land use conflicts into a uniform agricultural landscape.

Montgomery County also recognized that zoning imposed on the Agricultural Reserve only reserves it, and therefore does not fully relieve impermanence syndrome. The TDR program aimed to tackle that problem by motivating the permanent preservation of a critical mass of the Ag Reserve and consequently offering property owners belief in the long-term viability of agriculture here and confidence to make plans and investments accordingly.

Sending area owners must always retain at least one TDR per property of any size. They can also transfer just enough TDRs to be able to achieve the one-unit-per-25-acre maximum on-site density allowed by the rural density transfer zoning designation. In many or most cases, sending area property owners have elected to do just that and retain these usable TDRs. Because they permit on-site development of one unit per 25-acres, one out of every five TDRs became known as "super TDRs" with a much higher value than regular TDRs, prompting Montgomery County to later adopt a second Building Lot Termination program discussed below.

The allocation formula of one TDR per five areas turned out to sufficiently motivate participation from the owners of over 52,000 acres of sending site land. Significantly, Montgomery County has maintained the one-unit-per-25-acre zoning in the sending area since program inception in 1980.

Sending site owners sell their TDRs in private transactions to developers who want to exceed baseline density in receiving areas outside the Ag Reserve. At program startup, Montgomery County did not attempt to designate all of the receiving areas needed to absorb the entire sending area supply. TDR receiving areas were designated in master plan areas suitable for additional development due to location and infrastructure. Between 1981 and 2004, 82 subdivisions, or four percent of the total number of subdivisions approved in this period, used TDRs (Walls & McConnell 2007).

As of 2021, Section 4.9.17 of the zoning code indicated baseline and maximum densities for ten Rural Residential and Residential receiving zones. At the low end of this range, the RNC zone had a baseline of 0.2 units per acre and a maximum density with TDR of one unit per acre. At the high end, the R-10 zone had a baseline of 43.5 units per acre and a maximum with-TDR density of 100 units per acre. However, these are theoretical maximums that often are restricted by the master plans for each receiving area. For example, a 2007 study found that the R-200 zone theoretically allowed a maximum with-TDR density of 11 units per acre, (nine units over the baseline density of two units per acre), but on average the master plans allowed only three bonus lots per acre for a maximum with-TDR density of five units per acre (Walls & McConnell 2007).

When receiving site developments choose the TDR option, they must use at least 2/3 of the maximum number of development rights allowed in the respective TDR receiving zone unless the Planning Board finds that an exception should be made due to compatibility or environmental reasons.

About 70 percent of receiving site developments had met this 2/3 requirement as of 2007 and the other 30 percent had received exemptions (Walls & McConnell 2007).

Minimum densities required in developments using TDR must additionally fall within ranges for various building types which vary between the ten receiving zones. For example, in the RNC zone, all of the buildings must be single-family detached homes. In the with-TDR density range of 16 to 28 units per acre, detached, duplex, and townhouse units are not required to be any percent of the total number of units, but in developments with more the 200 total units, apartments must constitute at least 25 percent but cannot exceed 60 percent of the total number of units. The percent of the site which must be in common open space also increases from 0% in the RNC to 25 percent in all receiving site developments with a with-TDR density above 15 units per acre. As with the 2/3 rule, the Planning Board can grant exceptions to the building type mix and height limits based on environmental and compatibility features of the individual receiving site.

The number of bonus units allowed per TDR can be increased for receiving sites in two policy areas for specific building types. In a Non-Metro Station Policy Area, one TDR can yield two bonus apartment units. In a Metro Station Policy Area, one TDR can allow two bonus detached, duplex, or townhouse units or three bonus apartment units.

At receiving sites in Commercial/Residential and Employment zones, with-TDR residential bonus is expressed as FAR. Generally, each TDR yields 2,400 square feet of additional residential floor area. However, in a Metro Station Policy Area, each TDR allows an additional 4,400 square of residential floor area.

Montgomery County's TDR program worked well by observing eight of the ten factors found in those US TDR programs that have preserved the greatest amount of sending area land: Demand for bonus development; Customized receiving areas; Strict sending area regulations; Few alternatives to TDR; Market incentives that make TDRs attractive to buyers and sellers; Strong public support for preservation; Simplicity; and Promotion/Facilitation. Factor 10 is a TDR bank, a mechanism found in four of the top 20 US programs in 2009 but not in Montgomery County. In fact, Montgomery County formed a TDR bank in 1982 but eliminated it in 1990 because developers could easily find TDRs on the private market (Pruetz & Standridge 2009). Private transactions are greatly facilitated by real estate professionals who broker TDR sales.

Regarding Success Factor 6, a 2007 study found that Montgomery County did not provide developers certainty in the use of TDRs. Instead of allowing the use of TDR by right, Montgomery County, at least in 2007, negotiated final densities via an extensive process of development review (Walls & McConnell 2007).

The early success of this program was partly caused by Montgomery County's building boom in the 1980s that generated high demand for TDRs. TDR use at receiving sites peaked in 1983 at over 1,100 units and remained above 200 units per year for the rest of that decade. Receiving sites used less than 200 units per year in the 1990s, coinciding with a decline in the development of new subdivisions in general in Montgomery County in that decade (Walls & McConnell 2007).

The ability to use TDRs in suburban subdivision receiving sites was another reason for success. The receiving sites that used the bulk of the TDRs were in lower density zones including the RE-2 (baseline 0.4 units per acre, with-TDR density 1 unit per acre) and R-200 (baseline of two units per acre and with-TDR density of 11 units per acre). Although the R-200 allows a theoretical with-TDR density of 11 units per acre, the density actually achieved averaged five units per acre: two units baseline and three units per acre bonus using TDR (Walls & McConnell 2007).

Reasonable TDR prices also motivated TDR use in the early years of the program. TDR prices fluctuated from about $3,800 to roughly $7,500 between 1983 and 1990, but averaged approximately $5,000 per TDR. In the 1990s, prices ranged from $6,000 to $11,000 and averaged about $8,000 per TDR. By 2007, TDRs were selling for $20,000 on average, with some selling for as much as $45,000 each (Walls & McConnell 2007).

Building Lot Termination

Super TDRs, meaning those that allow development at the one-unit-per-25-acre density in the sending area, might be ten to 20 times more valuable than the regular TDRs that cannot be used in the sending area and only have value if they are transferred to a receiving area. In effect, this creates two separate TDR markets. In 2008, Montgomery County adopted a Building Lot Termination (BLT) program aimed at motivating sending site property owners to sell BLT easements that extinguish the right to build a dwelling unit on an eligible lot in the Ag Reserve.

Under BLT, eligible sending sites must meet seven requirements including: Agricultural Reserve zoning; at least 50 acres in size; at least 50% Class I – III soils or Woodland Classification 1 or 2; and capable of supporting an on-site waste disposal system. In addition, all regular TDRs on a sending site must be severed before or simultaneously with the termination of a building lot.

BLTs purchased by Montgomery County are retired but may be resold if no BLTs are available on the private market. Developers can also buy BLTs directly from owners of land in the Ag Reserve. BLTs increase density at receiving sites. BLTs are required to achieve incentive density on sites zoned Commercial Residential (CR) (59-C-15.87 and Life Science Center (LSC) (Chapter 59-C-5.473). BLTs can also gain public benefit points allowing increased density on receiving sites zoned CRT and EOF.

According to a 2016 presentation, receiving sites in the CR zone must use BLTs to achieve at least 7.5% of incentive density floor area, with each BLT yielding 31,500 square feet of incentive density and nine public benefit points. In the LSC zone, receiving sites must use BLT to gain 50% of the incentive density floor area above 0.5 FAR.

The County Executive annually sets the base value and maximum easement value that the county will pay for a BLT easement through its Agricultural Land Preservation Fund. The exact price paid by the county ranges between the base and the maximum price depending on the features of the specific property. These values are based on recent prices paid for TDRs, agricultural easements, BLT easements, and fee simple acquisition of agricultural land. In 2013, the base value was set at $222,390 (70 percent of FMV) and the maximum value was set at $254,160 (80% of FMV).

As of 2016, 41 BLTs had been created, representing the preservation of 1,025 acres of land in the Ag Reserve. As of March 2021, the Montgomery County website indicates that 61 BLTs had been created and 37 had been sent.

References
Pruetz & Standridge. 2009. What Makes TDR Work? Success Factors from Research and Practice. Journal of the American Planning Association. Vol. 75, No. 1, Winter 2009.

Walls, Margaret & Virginia McConnell. 2007. Transfer of Development Rights in U.S. Communities. Washington, D.C.: Resources for the Future.

Moraga, California, population 17,692 (2018), allows 1:1 transfers of density from hillsides designated as low-density residential or open space.

Morgan Hill, California, population 45,135 (2018), promotes preservation of steep hillsides with transfers at a 2:1 ratio from sending sites under county or city jurisdiction that allow city receiving sites to exceed baseline by up to ten percent. Projects using TDRs earn points that can improve their ranking under the city's annual building permit allotment system. The city has stocked its own TDR bank by severing credits from its hillside acquisitions and holding them for resale. Cash in lieu payments can also be used to gain additional receiving site density.

Mount Joy Township, Lancaster County, Pennsylvania, population 10,800 (2016) uses TDR to preserve farmland. TDRs produced by the preservation of sending sites zoned Agriculture are calculated by determining the number of lots that can be built on site in compliance with all codes and multiplying by five.

Mountain View, California, population 83,377 (2018), uses TDR in at least two precise plans to reduce density near natural areas, enhance habitat, and support the building of public schools. Under the San Antonio Precise Plan, purchase of a sending site for a public school was partly financed by the developers of six receiving sites to add bonus residential density and office floor area.

Mountlake Terrace, Washington, population 21,338 (2019), allows extra height to buildings in its downtown and Freeway/Tourist zone to developers who acquire TDRs from sending sites in the three-county Puget Sound Regional Program.

Narragansett, Rhode Island, population 15,504 (2017), allows development rights from wetlands, coastal resource overlay districts, areas within 200 feet of a blue line stream, and land within designated greenbelts to be transferred to receiving sites using its Planned Residential District procedures. Incentives that may be allowed include density increases of up to 20 percent.

New Castle County, Delaware, population 558,753 (2019), allows development rights to be severed from agricultural sending sites in the Suburban Reserve District and transferred to receiving sites in the Suburban District. Two receiving site homes can be built per development right from Class I soil and 1.3 homes per development right from Class II soil. For historic preservation, rights may be transferred between zoning districts within the same planning district. On receiving sites, transfers reduce open space requirements thereby increasing the net buildable area. As of 2012, TDR had preserved about 400 acres of farmland.

New Gloucester, Maine, population 5,803 (2018) uses TDR to preserve agriculture, forestry, and open space. Where the underlying zoning allows on-site development of one unit per five acres, the allocation rate is one TDR per two acres. Where the underlying zoning allows on-site development of one unit per two acres, the allocation rate is one TDR per one acre. Receiving area developments are allowed twice the density that would otherwise be allowed as long as the environment can accommodate the higher density.

New Jersey Highlands is an 859,358-acre region encompassing 88 municipalities and parts of seven counties. In 2004, the State of New Jersey adopted the Highlands Water Protection Act, which led to the 2008 Highlands Regional Master Plan and a TDR mechanism with a 398,000-acre sending area. Any of New Jersey's 218 municipalities can voluntarily create receiving areas for Highlands Development Credits in return for state grant funds to prepare necessary plans and codes plus the ability to charge impact fees on receiving area developments. While waiting for jurisdictions to participate, the New Jersey Highlands Development Credit Bank has purchased 1,814 credits representing the preservation of 2,565 acres.

New Jersey Pinelands, New Jersey, roughly one million acres in size, is managed under a regional plan implemented in part by TDR which aims to protect a critical aquifer as well as specialty agriculture, habitat, and other environmental resources. The entire profile is in Chapter 9: Water.

New York City, New York, population 8,419,000 (2019), was the first US city to use TDR and currently maintains two citywide transfer programs plus transfer mechanisms in ten special districts. Developers in New York City paid more than $1 billion for TDRs between 2003 and 2011. Despite a slump during the Great Recession, at least 421 development rights transactions occurred in New York during this nine-year period (Furman, 2013).

Historical Context – The historic landmarks TDR mechanism was adopted as part of the city's 1968 Landmarks Preservation Law. It allowed, and still allows unused density to be transferred from landmark buildings to adjacent sites in order to protect historic buildings from inappropriate alterations or demolition yet provide a form of compensation to the landmark owner for these restrictions. The city Landmarks Commission designated Grand Central Station as a landmark and denied the owner, Penn Central Transportation Company, permission to build an office tower on top of this building. The ensuing *Penn Central* lawsuit was ultimately decided in 1978 by the U.S. Supreme Court in favor of the city. The court found that the city's denial was not a regulatory taking under the Fifth Amendment of the US Constitution prohibiting governments from taking private property for public use without 'just compensation.' In its 1922 *Pennsylvania Coal Co. v Mahon* decision, the court found that a regulation can be a taking if it goes too far. But in *Penn Central*, the court did not conclude that a regulatory taking had occurred. Plus, the decision added: "…while these [development] rights may well not have constituted 'just compensation' if a taking had occurred, the rights nevertheless undoubtedly mitigate whatever financial burdens the law has imposed on the appellants and, for that reason, are to be taken into account in considering the impact of regulation." The US Supreme Court has yet to decide the question of whether or not TDRs constitute 'just compensation' in the event of an actual taking. However, the court's recognition of TDR as a legitimate mitigation measure gave communities throughout the US the confidence to adopt TDR programs.

Zoning Lot Merger (ZLM) - ZLMs treat contiguous tax lots as a single lot for zoning purposes, allowing the unused floor area from one tax lot to be freely used on another contiguous tax lot. The tax lots do not have to be under common ownership or lease. The mergers occur as a matter of right and require no approvals other than the recording of a Zoning Lot Development Agreement, known as a ZLDA or "Zelda".

ZLMs are not required to implement any particular planning goals. The Department of City Planning refers to ZLMs as "not technically TDRs" but feels obligated to treat them like TDRs since ZLMs operate in the same markets and produce results similar to TDR. A 2015 report from the Department entitled *A Survey of Transferable Development Rights Mechanisms in New York City* takes the position that it is problematic to place additional restrictions on ZLMs. "Tax lot lines reflect historic ownership patterns but typically do not relate to any land use purposes. Restrictions on the ability to merge them into unified zoning lots would give land use effect to tax lot lines, often without obvious underlying land use rationale. That may present legal and administrative difficulties" (NYC, 2015).

Although ZLMs are not required to implement planning goals, they sometimes do. Of the 421 transactions occurring between 2003 and 2011, 21 involved transfers from landmarks with 19 transfers resulting from ZLMs versus two under the Landmarks Program (Furman, 2013).

The original 1916 ZLM process allowed the construction of the Empire State Building in 1931. Changes were made to ZLM provisions in 1961, 1977 and 2001 after the Trump World Tower created controversy by concentrating development on 13 percent of a large, merged zoning lot. Today, ZLMs have facilitated several extremely tall towers in parts of Midtown that are not within contextual districts and consequently have no height limits. These ultra-high towers have drawn complaints from community organizations and the Municipal Arts Society, which published a critical report entitled "The Accidental Skyline" (NYC 2015).

Landmarks TDR Program: 74-79 – The Landmarks TDR program is often referred to as "74-79" for its section in the city Zoning Resolution. The 74-79 mechanism allows the difference between the floor area of a landmark and the floor area that would be allowed by zoning (if it were not a landmark) to be transferred to adjacent zoning lots. In 74-79, 'adjacency' means not just abutting lots but lots that are across streets and intersections and connected via chains of lots under common ownership. Despite this generous definition of adjacency, the Landmarks TDR program generated transfers from only 12 landmarks between 1968 and 2013 including Grand Central Terminal, Amster Yard, India House, John Street Methodist Church, Old Slip Police Station, 55 Wall Street, Rockefeller Center, Tiffany Building, Seagram Building, University Club, and St. Thomas Church for a total transfer of 1,994,137 square feet. An additional transfer between parts of Rockefeller Center was approved but not built (NYC 2015).

Observers cite the following reasons for reluctance to use 74-79.
1) The bonus on the receiving site is limited to 20 percent of the receiving site's as-of-right floor area.
2) The sending site is subject to a landmark maintenance plan.
3) The Special Permit needed for approval is time consuming and expensive (estimated by one expert to cost $750,000 (NYC, 2015).
4) Many landmarks have little or no unused floor area to transfer. Although New York has 1,300 landmarks, only 466 had unused development potential as of 2015 and many of these landmarks have so little unused potential floor area that transfers are functionally infeasible.
5) All three types of TDR mechanisms are hamstrung because TDR changes FAR limits while often not relaxing other development requirements, such as building height and setbacks, which also define the achievable building envelope. Developers can seek exceptions to these non-FAR requirements using the special permit or variance applications, but these processes are strict.
6) New York State Law limits residential buildings to 12 FAR, a scale that many developments can reach by incorporating affordable housing under the city's Inclusionary Housing Program.

Consequently, developers prefer whenever possible to use the ZLM process which is ministerial and therefore faster, cheaper, and more certain than 74-49. The numbers reflect this preference: between 2003 and 2011, 74-79 was used twice compared with 385 transfers occurring under the ZLM process (Furman 2013; Furman 2014; Gilmore 2013).

In its 2014 report, *Unlocking the Right to Build: Designing a More Flexible System for Transferring Development Rights*, the Furman Center argued for a retooling of the ZR 74-79 landmarks TDR program so that sending site owners have more potential users for their unused development rights. The report concludes that many benefits could result from this zoning code rewrite since the City had 1,400 designated landmarks as of 2014. In the portion of Manhattan south of Central Park alone, the report estimates that landmarks held 33 million square feet of unused development rights, enough transferable floor area to accommodate 33,000 additional residential units in receiving sites. Using TDR to create more housing units would relieve housing shortages and restrain rising housing costs that are forcing 47 percent of low-income renters to spend more than half of their income on rent and utilities (Furman 2014).

Theater District – In 1967, the city adopted a Special Theater District which used development bonuses that motivated the construction of five new theaters. However, the Theater District provisions were less successful at saving existing theaters. After losing several theaters, the city adopted a special permit process in 1982 that made it easier to transfer unused development potential from listed theaters and added a bonus for theater rehabilitation. However, the program was constrained by a limited receiving area. By 1988, the Theater Advisory Committee had grown the number of listed theaters to 30. To increase transfer options, the city allowed receiving sites in the Theater District outside the core to add one FAR of bulk in return for making a payment to preserve a theater within the core. This incentive proved inadequate. Between 1982 and 1998, only four transfers occurred yet theater owners had an inventory of over two million square feet of unused development potential (NYC 2015).

In 1998, the city adopted several reforms to its Theater District program.
- The receiving area was expanded, and listed theaters were allowed to transfer unused development potential anywhere in the Theater Subdistrict.
- The Special Permit requirement was replaced by a certification process allowing receiving area bonuses of either 20 percent or 44 percent FAR.
- Receiving sites were required to pay a surcharge with the proceeds used exclusively to fund theater preservation and use. Between 1998 and 2015, this surcharge increased from $10 to $17.60 per square foot of bonus floor area transferred to receiving sites.

To qualify for these bonuses, theater owners must covenant to maintain the theater for legitimate theater uses for the life of the receiving site development. TDR prices are determined by private transactions and in 2015 averaged about $225 per transferred square foot. Between 2001 and 2015, the Theater District became the city's most active TDR program, transferring roughly 500,000 square feet in 15 transfers from nine theaters (NYC 2015).

Grand Central Subdistrict – The 74-79 Landmarks Preservation TDR process transferred only 75,000 square feet from Grand Central Station between 1968 and 1992. In response, the city adopted the Grand Central Subdistrict of the Special Midtown District in 1992 to further the preservation of the Grand Central Station and other landmarks plus maintain pedestrianism and area character. In

1992, Grand Central Station was the only designated landmark with any appreciable amount of unused development potential, but it had a 1.7 million square feet of TDR supply. To motivate transfers, the 1992 changes allowed sending site TDRs to be transferred to receiving sites anywhere in the subdistrict to achieve one FAR of bonus floor area via a certification process. Also, by special permit, receiving sites in this subdistrict's core can achieve up to 21.6 FAR. Between 1992 and 2015, five transfers from Grand Central Station increased the total transfer from Grand Central Station to 488,036 square feet (NYC 2015).

Special West Chelsea District – New York wanted to convert an abandoned elevated railway into the High Line pedestrian open space now beloved by residents and tourists. In 2005, New York adopted the Special West Chelsea District which allowed owners of properties constrained by the High Line to transfer their unusable development rights to receiving areas in most of the district's nine subdistricts. Importantly, the receiving sub-districts were not adjacent to the High Line due to the desire to create light, air and views from the High Line. Transfers are allowed by notification, a process that is easier than certification and considered the least onerous mechanism of any TDR transfer process allowed in New York City. The prices for these TDRs have ranged from $200 to $400 per square foot. The receiving site bonuses are in tiers that require a certain number of TDRs to be used before other bonuses can be added in order to achieve maximum development potential. For the non-TDR tiers, density bonus is granted for affordable housing and restoration of the High Line itself. As of 2015, 403,983 square feet of development had been transferred from 25 sending sites (NYC 2015).

South Street Seaport – The city created the South Street Seaport Subdistrict in 1972 to support the South Street Seaport Museum plus preserve and restore the historic Schermerhorn Row landmark buildings which were defaulting on mortgages and at risk of demolition. A TDR bank was created to immediately save the buildings and hold the TDRs for later use on designated South Street commercial area receiving site developments. Chase Manhattan Bank and Citibank accepted the TDRs as partial satisfaction of loan obligations. In contrast with most traditional historic preservation TDR programs, the allocation formula is the floor area of the sending site building or five times the lot area including the land area of adjacent abandoned streets. Transfers were conducted by the relatively easy process of certification. By 2013, all but 340,000 square feet out of a total supply of 1.4 million square feet had been transferred to six receiving site projects at prices ranging from $110 to $150 per square foot in 2007-2008 (NYC 2015).

Hudson Yards Special District – In 2005, the city adopted a special district which included two TDR mechanisms to help create an open space network without the expense of condemnation or acquisition as part of the redevelopment of the Metropolitan Transportation Authority's rail terminal facilities known as Hudson Yards. Like the West Chelsea Special District, maximum density in some of the receiving areas can only be achieved by combining TDRs with other bonus options. MTA owns the development rights and established the TDR price at 65 percent of the receiving site's appraised development rights, which in 2015 was about $350 per square foot. As of 2015, the transfer of 566,628 square feet of TDRs was pending (NYC 2015).

Special Manhattanville Mixed Use District – This district, adopted in 2007, allows Columbia University to move density between sending and receiving sites owned by the University in order to create open space in a new campus while retaining its full development potential (NYC 2015).

Sheepshead Bay Special District – This TDR program was created in 1973 to promote preservation of a revered restaurant by allowing transfers within a 20-block district. The program allowed a TDR bank to facilitate transfers but the restaurant closed and the district had not experienced any transfers by 2015 (NYC 2015).

Coney Island Special District – The city adopted the Coney Island Special District in 2009 to preserve a designated landmark restaurant and promote general revitalization by allowing transfers as a matter of right within the district. In 2012, Hurricane Sandy forced the city to reevaluate these plans (NYC 2015).

United Nations Special District – In 1970, the city adopted a special district that included a mechanism to move bulk around within the area surrounding the United Nations complex (NYC 2015).

Hudson River Park Special Transfer District – In 2015, the city approved a mechanism to rehabilitate Pier 40 in Hudson River Park by severing and selling TDRs to a large mixed-use development with senior and affordable housing on the other side of the West Side Highway (NYC 2015).

References

Furman Center for Real Estate & Urban Policy. 2013. Buying Sky: The Market for Transferable Development Rights in New York City. New York: New York University.

Furman Center for Real Estate & Urban Policy. 2014. Unlocking the Right to Build: Designing a More Flexible System for Transferring Development Rights. New York: New York University.

Gilmore, Kate. 2013. A Process Evaluation of New York City's Zoning Resolution (ZR) 74-79: Why Is It Being Used So Infrequently? New York: Columbia University.

New York City Department of City Planning (NYC). 2015. A Survey of Transferrable Development Mechanisms in New York City. New York: Department of City Planning.

Newport Beach, California, population 84,534 (2019), allows transfers between properties within the same Statistical Area of the General Plan that reduce traffic, improve development character, make more efficient use of land, or preserve historic buildings, natural resources, view corridors, and open space.

Normandy Park, Washington, population 6,604 (2019), uses a TDR mechanism in its Manhattan Village neighborhood aimed at inter-jurisdictionally transferring TDRs from sending sites on Vashon Island in King County via interlocal agreement. The program promotes mixed use development on receiving sites in an effort to improve sale tax revenues and community vitality.

North Kingston, Rhode Island, population 26,320 (2019), uses TDR to preserve sensitive resources including groundwater reserves, wildlife habitat, agricultural land and access to surface water. In three receiving-site districts, TDR allows bonus residential density, non-residential intensity or mixed uses.

Northbrook, Illinois, population 32,958 (2019), uses TDR to shift development away from a closed landfill to more suitable locations within the village's Techny Overlay District. One transfer relocated 1.5 million square feet of floor area.

Northampton, Massachusetts, population 28,726 (2018), allows transfers from its Farms, Forests & Rivers Overlay to the Planned Village Development Overlay.

Oakland, California, population 429,082 (2018), allows transfers of density and intensity between adjacent sites for historic preservation and other city goals.

Ocean City, Maryland, population 6,972 (2019), is a coastal resort town on a barrier island along Maryland's Atlantic shore which uses TDR to reduce vulnerability to coastal storms exacerbated by sea level rise. The complete profile is in Chapter 6 on sea level rise in Part I.

Okeechobee County, Florida, population 42,168 (2019), uses TDR to preserve wetlands and uplands containing critical habitat.

Orange County, Florida, population 1,393,000 (2019), surrounds Orlando and is home to retirement communities, golf resorts and tourist attractions including Universal Studios and Walt Disney World. In 1997, the county adopted a TDR program aimed at promoting conservation in conjunction with development of Horizon West, a 23,000-acre area planned for mixed use villages and a town center, each surrounded by green space. The property owners here partnered as Horizon West, Inc. and joined with Orange County to develop the Horizon West plan.

As it appears in January 2021, Orange County's Article XIV, Division 3 Transfer of Development Rights in the Village Land Use Classification states that the TDR program is one tool, but not the only tool, designed to preserve agricultural land, natural resources, and green space. A specific area plan (SAP) designates sending areas (including greenbelts and wetlands), receiving areas, overall village density and open space. Transfer of density within a village must conform with the SAP.

Sending areas include wetlands and upland greenbelts, a term that includes green space and wildlife corridors within a village as well as perimeter open space intended to establish each village as a separate and unique place. TDR allocation ratios vary between the five villages and one town center. In Lakeside Village, upland greenbelts can generate 11 TDRs per acre and wetlands can produce one TDR per 3.5 acres preserved by easements or covenants. In Bridgewater Village, upland greenbelts can generate 17.1 TDRs per acre and wetlands can produce one TDR per 2.9 acres.

Receiving areas are in village centers and other areas established by each SAP. In 2021, the code included 16 density tiers with baselines ranging from one unit per acre to 25 units per acre. Unlike most TDR programs, this ordinance requires TDRs to reduce as well as increase density. For example, in a tier with a baseline density of three units per acre, a 20-acre site could develop 60 units without TDR. However, 11 TDRs would be needed to increase density to 71 units or reduce baseline density to 49 units.

Deviation from baseline density differs depending on whether the receiving site development is exceeding or falling short of baseline density as well as which density tiers applies to the receiving site. In a tier where baseline is two units per acre, TDR can double density to four units per acre. But in a tier with a baseline density of 12 units per acre, TDR can only add eight units per acre for a maximum density of 20 units per acre. In the highest density tier, baseline is 25 units per acre and TDR cannot be used to exceed baseline. Each development that uses TDR to exceed baseline must prepare a report demonstrating that the proposed transfer and all cumulative transfers will not cause the maximum capacity of the receiving site's neighborhood school to be exceeded.

A different set of calculations apply to receiving site projects proposing lower than baseline density. For example, in a tier where baseline is two units per acre, TDR can reduce density to no less than one unit per acre, a 50 percent decrease. But in a tier with a baseline density of 20 units per acre, TDR can reduce density to no less than 16 units per acre, a 20 percent decrease. Each development that uses TDR to exceed baseline must prepare a report demonstrating that the proposed transfer and all cumulative transfers will not cause the attendance at the neighborhood school to drop below its minimum enrollment.

Over time the county has amended this code section by adding additional tiers with higher baseline densities. In 2012, the code included five receiving site categories with baselines of two units per acre on the low end and 12 units per acre on the high end of the density range. As noted above, the code now has 16 density tiers topping out at a tier with a baseline density of 25 units per acre, a density that would not be achievable in 2012 even with the use of TDR.

The county prepared an assessment of the Horizon West planning area in 2012. At that time, 75 percent of the planning area was still used for agriculture or characterized by vacant parcels under 270 different ownerships. By 2012, 5,810 dwelling units had been built, with about half in single-family residences and another 25 percent in townhouses. Some subdivisions approved for construction by various developers were only partially completed or not started at all. The assessment primarily attributes the slow progress to the real estate downturn caused be the Great Recession.

The 2012 assessment reported that the TDR program had been hampered by regulatory changes and market preferences. For example, the code was changed to eliminate the requirement to use TDR for townhouses and the assessment notes a past market preference for larger lots. The assessment also observed that the demand for TDRs is limited by the capacity of each neighborhood school. The report states that 1,904 TDRs were "accumulated" in Horizon West as of 2006 (Orange County 2012). Linkous and Chapin (2014) reported no TDR activity as of 2014.

References

Orange County. 2012. The Horizon West Retrospective: An Assessment of Florida's First Sector Plan. Accessed 2-5-21 at HW_Retro_Report_April_6_2012.pub (orangecountyfl.net).

Linkous, E. and T. Chapin. 2014. TDR Program Performance in Florida. Journal of the American Planning Association, 80:3, 253-267.

Osceola County, Florida, population 375,751 (2019), uses TDR to protect agricultural land, preserve historic heritage and reduce/eliminate incompatible uses near military facilities. Receiving areas include unincorporated land within the Urban Growth Boundary and sites established through interlocal agreement.

Oxnard, California, population 209,877 (2018), uses TDR to motivate the redirection of development away from significant resources, hazardous areas and places of potential public access on the coast. Beachfront sending sites can be issued up to six TDRs per lot and potential receiving sites lie within four coastal and three non-coastal zones. Receiving site developers can choose to pay cash in lieu of actual TDRs and Oxnard must use the proceeds to buy beachfront lots. Oxnard exempts extra units added to receiving sites via TDR from plan check fees, park fees, and growth development fees. Some transfers occurred in the 1980s. But since then, the TDR option has not motivated owners to forego development potential on valuable oceanfront property.

Pacifica, California, population 38,984 (2019), uses TDR to preserve areas subject to landslides, flooding and other hazards plus sending sites designated by the Open Space Inventory, the Planning Commission and the City Council. Receiving areas are multi-family residential zones and planned developments.

Palm Beach County, Florida, population 1,497,000 (2019), is located on the Atlantic Coast roughly 60 miles north of Miami. The county operates a TDR program using a bank that sells TDRs severed from 34 nature preserves consisting of 31,000 acres originally preserved by voter-approved bond measures. The profile for this program appears in Chapter 8 on biodiversity in Part I.

Palmer, Massachusetts, population 12,232 (2019), employs TDR to preserve rural, historic and agricultural character. Receiving sites can use TDR to increase residential density, non-residential floor area, lot coverage, or add a neighborhood commercial building lot in a Traditional Neighborhood Development. Receiving site developers can comply using actual TDRs or a cash in lieu amount determined by the Conservation Commission.

Palo Alto, California, population 66,573 (2019), lies on the western shores of San Francisco Bay, 30 miles southeast of downtown San Francisco and 15 miles northwest of downtown San Jose. It is the home of Stanford University and the headquarters of several high-tech companies including Hewlett-Packard.

Palo Alto uses various incentives to protect its historic resources, including a TDR program that encourages property owners to perform seismic and historic rehabilitation in the City's Downtown Commercial District (DCD). The program also allows qualified City-owned historic properties located in any zoning district to transfer TDRs to eligible receiving sites in the Downtown Commercial District.

The DCD portion of Palo Alto's zoning code creates five categories of floor area bonuses. One of these floor area bonuses, not discussed below, cannot be transferred to a qualified receiving site. Of the remaining four categories, the first three bonuses discussed below may be transferred to a qualified receiving site or used on site at the owner's option. Bonuses gained in the fourth category, Historic Bonus for Over-Sized Buildings, must be transferred off-site to a qualified receiving site.

- Seismic Rehabilitation Bonus - Buildings in Seismic Categories I, II or III that undergo seismic rehabilitation may receive a bonus of up to 2,500 square feet or 25 percent of existing floor area (whichever is greater). (This option cannot be used on buildings that exceed specified FAR limits in specified districts.)
- Historic Rehabilitation Bonus - Buildings in Historic Categories 1 or 2 that undergo historic rehabilitation may receive a bonus of up to 2,500 square feet or 25 percent of existing floor area (whichever is greater). (Again, this option cannot be used on buildings that exceed

specified FAR limits in specified districts.)

- Combined Historic and Seismic Rehabilitation Bonus - Buildings in Historic Category 1 or 2 undergoing historic rehabilitation which are also in Seismic Categories I, II or III and undergoing seismic rehabilitation may receive a bonus of up to 5,000 square feet or 50 percent of existing floor area (whichever is greater) . (Once again, this option cannot be used on buildings that exceed specified FAR limits in specified districts.)

- Historic Bonus for Oversized Buildings – Buildings in Historic Category 1 or 2 undergoing historic rehabilitation that exceed FAR 3.0 (or FAR 2.0 in specified sub-districts) can obtain a floor area bonus of 50 percent above the maximum that would otherwise apply within the applicable sub-district. As mentioned above, this bonus can only be used by transferring it to a qualified receiving site.

When a City-owned property is proposed, the City Council must designate the property as an eligible sending site. The property must be in Historic Category 1 or 2 and/or in Seismic Categories I, II or III. Concurrent with designation of the sending site, the City Manager must create a special fund for the proceeds of the TDR sales. This money must be committed to the rehabilitation of the sending site property or any other city-owned property in Historic Category 1 or 2 and/or in Seismic Categories I, II or III.

To be certified as a receiving site, a property must meet all of the following criteria.

- Within the Downtown Commercial District
- Not a historic site or a site with a historic structure
- Various requirements for distance from residential zones

When receiving sites are outside the downtown parking assessment district, the additional development allowed through TDR is limited to floor area ratio of 0.5 or 10,000 square feet of additional floor area, whichever is greater. When receiving sites are within the downtown parking assessment district, the additional development allowed through TDR is limited to a floor area ratio of 1.0 or 10,000 square feet of additional floor area, whichever is greater. In no event can TDR transfers cause the site to exceed FAR 3.0 in the CD-C sub-district or FAR 2.0 in the CD-S and CD-N sub-districts.

Transfers can be made directly to receiving sites or to intermediaries as long as the conveyance is evidenced by a recorded document that identifies the sending site.

In 2007, the city sold a 2,500 square foot bonus generated by the rehabilitation of the Children's Library for $237,500 or $95 per square foot. In 2009, the city approved sending site status for the former Sea Scout Building, a Category 1 historic building designed by Birge Clark and built in 1941. That same year, the city council also approved eligibility for the city to sell 2,500 square feet of bonus floor area for the rehabilitation of a 1936 building occupied by the College Terrace Library and Palo Alto Child Care Center.

In 2015, the city approved eligibility to sell TDRs for the rehabilitation of the city-owned 1932 Roth Building, listed in the National Register of Historic Places. This building has long been considered an ideal site for a city history museum. However, a 2020 news item indicated that rehabilitation work had not started because the funds needed for the work, including $4.9 million expected to be generated from TDR sales, was insufficient to complete an estimated $6 million to $10.5 million in repairs and seismic upgrades.

In 2018, the city council approved the sale of TDRs for renovation of the city owned Avenidas Building, a historic Spanish Colonial Revival building constructed in 1927.

Park City, Utah, population 8,296 (2018), uses TDR to preserve open space, forests, agriculture, scenic views, environmental areas and historic properties. In receiving areas, one development credit adds 1,000 square feet of commercial floor area or 2,000 square feet of residential floor area.

Pasadena, California, population 141,258 (2019), adopted a TDR program to mitigate the impact of the city's 1985 Central District Plan. The original program preserved at least two historic landmarks before being eliminated in 2005. However, TDR remains intact in Pasadena's West Gateway Specific Plan and has saved historic gardens and buildings including Ambassador Auditorium.

Pass Christian, Mississippi, population 6,307 (2019), has suffered extensive hurricane damage and adopted a Smart Code aimed at motivating the permanent preservation of a transect consisting of floodplains, corridors, steep slopes, woodlands, farmland, and view sheds as well as open space, corridors and buffers planned for acquisition. The sending area allocation ratio is one Equivalent Housing Unit (EHU) per acre of land preserved. Each EHU can be used to build one dwelling unit, two motel/hotel rooms or 1,000 square feet of non-residential floor area. In the lowest density of the four zones that serve as receiving areas, baseline is 12 EHUs per acre and maximum density/intensity is 18 EHUs per acre. In the highest density zone, baseline is 36 EHUs per acre and maximum density/intensity is 75 EHUs per acre.

Payette County, Idaho, population 23,951 (2019), uses TDR to preserve agricultural land, habitat and open space. Upon the transfer of a development right, the sending site is permanently precluded from receiving TDRs. However, according to Section 8-5-10.F: "This disqualification shall not prohibit any landowner from later applying for a rezone and subdivision approval after fifty (50) years from the approval of any application to transfer a development right." As of 2009, this program had protected 4,145 acres.

Peach Bottom Township, York County, Pennsylvania, population 5,050 (2018), uses TDR to preserve farmland by allowing development rights to be transferred to receiving sites with lower agricultural value.

Penn Township, Lancaster County, Pennsylvania, population 9,870 (2018), uses TDR to preserve farms, natural areas and rural character. Receiving sites can use TDRs to build above baseline density, build below baseline density, plus exceed baselines for building footprint, lot coverage and building height.

Perinton, New York, population 46,713 (2016), uses TDR to preserve nature, open space, and recreational lands by transferring development at a 1:1 ratio.

Pierce County, Washington, population 904,980 (2019), stretches from the west side of Puget Sound 60 miles east into the Cascade Mountain Range. The center of the County includes Tacoma and other rapidly growing urban areas in the greater Seattle region. But the western side of Pierce

County consists largely of lightly populated peninsulas and islands while the eastern half features Mount Rainier National Park, the Mt. Baker/Snoqualmie National Forest, and several wilderness areas. Despite the relatively high percentage of preserved land, Pierce County and many of its incorporated cities have preserved additional land using multiple approaches including TDR.

In 2007, the County adopted a TDR program that conserved 90 acres in its first two transactions in 2009. In 2011, the county established administrative procedures for tracking and banking TDRs. In 2012, the county preserved 120 acres of farmland and used the resulting 73 TDRs as the initial deposit in its TDR bank. The Pierce County Council adopted revisions to its TDR ordinance in 2013, 2016, and 2017. The program description below uses the code that existed in October 2020.

Pierce County's TDR code aims to permanently conserve resource lands, including forestry, agriculture, trails, habitat, open space and rural land in general. Some land is eligible to become sending sites based on its zoning as Agricultural Resource Land (ARL) or Forest Lands (FL). Properties can also qualify by being in rural zoning districts, identified for trail development, determined as having habitat for endangered or threatened species, or specified within an interlocal agreement. Land within urban growth areas or designated as urban or rural shoreline can also qualify if it is threatened with development within ten years, included in an inventory of open space, and providing recreational activities, such as golf and soccer, that do not require intensive development.

The conservation easements needed to sever TDRs from sending sites generally allow uses consistent with the purpose of its site's zoning. For example, the easements placed on sending sites zoned ARL permit agricultural uses consistent in that zone. In addition, conservation easements placed on sending sites zoned FL must be accompanied by an approved forest stewardship plan. The easement must also identify any development rights that the sending site owner chooses to retain for future development.

Generally, the number of TDRs available is determined by the on-site development allowed under the site's zoning. However, sending sites zoned ARL within two community planning areas as well as urban growth boundaries can yield one TDR per acre rather than one TDR per five acres. Likewise, in some areas, the maximum density allowed under the previous zoning allows an allocation of six TDRs per acre rather than the one TDR per five-acre allocation that would typically apply to land zoned ARL.

TDRs can be transferred to receiving sites in other jurisdictions in accordance with an executed interlocal agreement. The City of Tacoma and Pierce County signed an interlocal agreement in 2012. As detailed in the profile of Tacoma, that city uses TDR to protect city landmarks, preserve habitat within the city, and conserve resource lands in King and Snohomish counties as well as preserve farmland under Pierce County jurisdiction in the Puyallup Valley and forest land zoned FL throughout Pierce County. The Pierce County TDR Technical Oversight Committee makes recommendations on all interjurisdictional transfer proposals. Interjurisdictional transfers are not allowed if the receiving site's jurisdiction objects.

Land under Pierce County jurisdiction can become TDR receiving sites if part of an urban project or comprehensive plan amendment involving increased density. Land proposed as an urban growth area expansion and annexation are guided by an interlocal agreement between Pierce County and the affected city or town.

The number of bonus dwelling units allowed by TDR is calculated by subtracting the original (pre-transfer) units from the final (post transfer) units and dividing by a conversion factor specified for each of 25 different final density categories ranging from one unit per acre to 25 units per acre. The conversion rate increases as the final density increases. For example, a one-acre site with an original density of four units per acre and a final density of 14 units per acre would require two TDRs because the difference of ten bonus units is divided by five, the conversion rate for the 14-unit per acre final density category. The conversion rate tops out at eight for projects within the 25-unit per acre density category.

The TDR Technical Oversight Committee can lower the required number of TDRs if a receiving site development has made or is planning to make significant infrastructure improvements near the receiving site. Conversely, the TDR Technical Oversight Committee can increase the required number of TDRs if a receiving site development has been relieved from making infrastructure improvements because the state or county has made or is planning to make significant infrastructure improvements near the receiving site.

Developers can achieve bonus development potential on receiving sites by buying TDRs from sending site owners or the Pierce County TDR Bank. If the bank does not have sufficient credits in stock, developers may also provide funds for the bank to buy the necessary TDRs at a cost determined by the TDR Administrator in consultation with the TDR Technical Oversight Committee.

The bank sells TDRs for the current fair market value as set by the TDR Technical Oversight Committee. The bank uses TDR sale proceeds to facilitate the TDR program, including executing TDR purchases and sales, retiring debt incurred by PDR programs, providing amenity funding pursuant to interlocal agreements, marketing, procuring appraisals, and reimbursing governments for program administration.

The bank cannot buy TDRs for more than fair market value as determined by an appraisal that subtracts the value of the land with the rights removed from the pre-transfer value of the land.

Pierce County can motivate cities and towns to accept TDRs by offering amenity funds, typically used for additional transit facilities, parks, road/sidewalk improvements, public art, and community facilities. The interlocal agreement controls the type, amount and timing of amenity funding.

In 2016, the program accomplished its first interjurisdictional transaction which saved farmland in Pierce County and allowed an increase of 21 units to the Stadium Apartments receiving site in the City of Tacoma. As of January 2021, the Pierce County TDR program had saved approximately 375 acres of farmland.

Pima County, Arizona, population 1,047,279 (2019), uses TDR to preserve habitat, land near military airports, floodplains, geologic features, recreation areas, and land with unique aesthetic, architectural or historic value. Receiving areas are mapped and developments here can be approved without rezoning.

Pismo Beach, California, population 8,213 (2018), uses TDR to retain open space, preserve scenic views, reduce hazards, and/or provide public access by allowing 1:1 transfers from sending sites not precluded by development constraints, such as steep slopes, or within 25 feet of the edge of a coastal bluff.

Pitkin County, Colorado, population 17,767 (2019), surrounds the affluent ski resort of Aspen in the mountainous center of the state. To keep growth from degrading environmental resources, the county aims to steer higher-density development into the Urban Growth Boundary (UGB) and allow mostly lower-density residential, agriculture/ranching, and open space outside the UGB. The county also anticipates that climate change will cause wildfires to continue increasing in size, intensity and frequency over time. To reduce human exposure to fire and other disasters, the county's Hazard

Mitigation Plan recognizes that the TDR program works with UGB policy by motivating the redirection of growth from hazard-prone areas and rural zones into places more suitable for development where it can benefit from infrastructure, emergency services, and public services.

In 1996, the Pitkin County Board created the county's Rural/Remote (R/R) zoning district to limit growth in backcountry areas which are characterized by sparse development, little or no access to traditional county emergency services, absence of utility districts, and exposure to natural hazards. This zone requires a minimum lot size of 35 acres and cabins here cannot exceed 1,000 square feet of floor area. Alternatively, owners of property in the R/R can choose to record covenants permanently precluding further development in return for one TDR per each full 35 acres or legally created parcel smaller than 35 acres. Concurrent with the recording of the covenant, the county issues an irrevocable certificate of development rights. Within five days of the sale, assignment, conveyance, or other transfer or change of ownership of a TDR certificate, the county must be notified of the certificate number, the grantor, the grantee, and the total value of the consideration paid for the certificate.

Since then, Pitkin County expanded opportunities for owners to participate by adding sending area potential for various forms of preservation in other rural districts including environmentally sensitive properties and land with environmental hazards including geologic instability, steep slopes, and vulnerability to wildfire.

As of January 2021, sending sites include the following:
- Legal lots and parcels zoned R/R or Transitional Residential (TR-1 and TR-2);
- Land zoned Conservation Development PUD (CD-PUD);
- Constrained or visually-constrained land not zoned R/R, CD-PUD, TR-1 or TR-2;
- Limited Development Conservation Parcels zoned AR-10, RS-20, RS-30, RS-35 or RS-160; and
- Properties designated in the Pitkin County Historic Register.

As of January 2021, the following allocation ratios apply:
- The allocation ratio in the TR-1 is the same as in the R/R: one TDR per 35 full acres or legal lot larger than one acre but smaller than 35 acres. Under the conditions described below, a legal lot zoned R/R or TR-1 that is smaller than one acre may also qualify for one TDR;
- In the TR-2 zone, the code allows one TDR per ten full acres or legal lot larger than one acre but smaller than ten acres. Under some circumstances a legal lot zoned TR-2 (as well as R/R and TR-1) that is smaller than one acre may qualify for one TDR by proving that the site has legal access and can accommodate a well, on-site wastewater treatment and a 1,000 square foot building pad;
- At a public hearing, the Board of County Commissioners (BOCC) may issue one or more TDRs to constrained sites or one TDR to visually constrained sites;
- For sites zoned CD-PUD, one TDR can be issued for each 35 acres preserved and excluding the 160 acres designated for development);
- On a Limited Development Conservation Parcel, one TDR can be issued for each 35 acres preserved after deducting 35 acres for each legal dwelling on a parcel less than 640 acres and deducting 70 acres for each legal dwelling on properties with 640 acres or more; and

- The BOCC may approve the granting of TDRs to properties designated in the Pitkin County Historic Register.

The Pitkin County TDR program motivates owners of qualified sending sites to participate with sale prices that ranged from $115,000 to $318,000 per TDR between 2007 and 2019. In 2019, ten TDRs sold at prices ranging between $225,000 and $240,000 each (Condon 2020).

TDRs cannot be received on land zoned R/R, RS-160, TR-1, RS-G, MHP, AH, AH-PUD, B-1, B-2, VC, P-I, T, SKI-REC, VR, I, PUD, AC-REC-2 or FPV-O.

In receiving areas, property owners can use TDRs from any sending site to obtain an exemption for a new development right in the Aspen UGB under the Growth Management Quota System (GMQS). The GMQS uses a point system to score proposed developments based on how well they meet desired criteria. A proposed development that scores high may receive a permit to build a residential unit and/or gain bonus floor area within an individual dwelling unit through the GMQS annual quota without having to buy TDRs. However, a low-scoring project may have to wait year after year unless the developer uses the TDR option. The number of TDRs required for exemptions from GMQS varies depending on the size of the proposed residence. Transfers used to receive an exemption to build a new residential unit are approved by a One-Step Special Review of the BOCC.

TDRs from any sending site can also be used to exceed a baseline floor area of 5,750 square feet on receiving sites in the Aspen UGB and approved subdivisions up to the final maximum floor area allowed within each individual home. For receiving sites zoned TR-2, each TDR yields 1,000 square feet of additional floor area. On receiving sites in other zones, the ratio is 2,500 additional square feet of floor area per TDR. These transfers are approved by a One-Step Special Review conducted by the Hearing Officer.

As of 2020, the county had issued certificates for 389 TDRs and 254 TDRs had been transferred to receiving sites, with 82 percent of these TDRs used to exceed the baseline floor area of 5,750 square feet. At that time, TDRs had preserved a total of 8,879 acres.

In October 2020, county staff estimated that 20 years would be needed to absorb the remaining 135 TDRs unless the program was modified. At that time, staff was considering various ways of dealing with the possibility of TDR oversupply, including the cessation of issuing new TDRs. However, a real estate agent who assists with TDR transactions argued that the county should not stop issuing new TDRs because the apparent oversupply of existing TDRs was misleading. The agent estimated that 12 TDRs had been sold in the prior year and that he was having difficulty finding the ten TDRs wanted at receiving sites because many of the 135 unextinguished TDRs were "locked up" and could not be used in the near term for various reasons. As of October 2020, staff planned to expand public outreach as part its review of all county growth management tools including TDR (Condon 2020).

Reference

Condon, S. 2020. Transferable development rights under scrutiny as Pitkin County sharpens growth management tools. Aspen Times: 10-10-20. Accessed 2-19-21 at Transferable development rights under scrutiny as Pitkin County sharpens growth management tools | AspenTimes.com.

Pittsburgh, Pennsylvania, population 302,205 (2019), uses TDR in its downtown to preserve historic landmarks and performing arts centers.

Pittsford (Town), Monroe County, New York, population 29,265 (2018), has incentive zoning provisions that allow developers to deviate from various development requirements including land use changes, reduced roadway standards, modified lot dimensions, and density bonus in two receiving site zones. In return, the developers propose any of a number of benefits including the construction of community facilities or the preservation of landmarks, open space, ecological areas, and farmland.

Plymouth, Massachusetts, population 60,803 (2018) uses TDR to protect wellheads, aquifer recharge zones, public water supply, environmental areas, access to natural resources and land with potential municipal use. To adjust for diverse sending area values, TDR allocation is calculated by the formula FMV/AVG. FMV is the difference between the appraised value of the lot sales and infrastructure costs. AVG is the average assessed value of a buildable lot in the RR zone. In two receiving zones, the use of TDR requires a special permit process. Developers can get permission for TDR projects from the planning board using only the site plan approval process when RR receiving sites are at least five acres in size and have at least 500 feet of frontage on a major street.

Pocopson Township, Chester County, Pennsylvania, population 4,840 (2018), uses TDR to preserve farmland, natural areas, historic resources, and rural character. The additional development allowed on receiving sites varies depending on the nature of the sending site, such as a woodland or greenway.

Polk County, Florida, population 724,777 (2019), allows development rights to be transferred between properties no more than one mile apart as long as the receiving site does not have a lower permitted density than the sending site. Development rights can be converted from one land use to another based on vehicle trip generation rates. Over 200 TDRs have transferred.

Port Orchard, Washington, population 14,062 (2019), has an interjurisdictional program that preserves farms, open space, and rural character on sending sites in Kitsap County with transfers to city receiving sites where each TDR allows 2,000 square feet of a one-story height bonus over baseline height limits.

Port Richey, Florida, population 2,831 (2019), allows transfers of residential density and non-residential intensity to preserve environmentally sensitive land.

Portland, Oregon, population 645,291 (2019), uses TDR in several planning districts to achieve various goals including the preservation of housing, open space, viewsheds, riparian corridors, habitat, environmental resources, historic landmarks, infrastructure capacity, and trees.

<u>Northwest Hills</u> – Portland adopted its first TDR program in 1991 as an implementation tool for its Northwest Hills Plan. The planning area includes a wooded bluff on the west side of the Willamette River that creates a natural backdrop for seven miles of the city center. The plan aims to maintain the character and natural resources here by limiting development within an Environmental Protection overlay where land zoned for single-family residential dwellings can become TDR sending sites by reducing or eliminating on-site development by covenant.

The density is transferred at a one-to-one ratio by planned unit development applications submitted for the sending and receiving sites. Receiving sites are zoned Residential Farm/Forest (RF) that are within the Northwest Hills district but not covered by the Environmental Protection overlay. Baseline in the RF zone is one unit per two acres. Transfers boost density to one unit per acre for qualified sites.

Johnson Creek – In 1996, Portland adopted the Johnson Creek Basin Plan aimed at shifting development away from sensitive environmental areas and locations vulnerable to flooding. Sending sites can transfer the density allowed by the underlying zoning to any property in the Johnson Creek district that is not zoned Residential Manufactured Dwelling Park and not within selected subdistricts, restricted by the Environmental Protection overlay, or within a flood hazard area. Transfers are processed as planned development applications for both sending and receiving sites. TDR can increase development on receiving sites up to 200 percent of baseline density.

In competition with TDR, developers here can also gain density by building attached units that meet specifications for sewer, water, and storm water management located within ¼ mile of a transit street or transit way. In addition, the incentive to use TDR in the Johnson Creek district has been reduced by programs that buy out the owners of flood prone properties.

Landmarks – Portland encourages the preservation of historic landmarks using procedures that vary depending on the zoning districts involved. For example, in multi-dwelling residential zones, sending sites with qualifying historic resources can transfer unused FAR plus 50 percent of the maximum FAR allowed by the site's zoning if the building is seismically safe, or the owner executes an agreement to meet seismic safety standards. Sites in any of nine zones can receive transferred FAR on a one-to-one basis up to the maximum FAR increase specified in the code for all types of bonuses, which varies between zones.

Affordable Housing – Portland uses TDR to motivate the preservation of housing affordable to households earning no more than 60 percent of area median family income. In the multi-dwelling residential zones, sending sites with affordable housing can transfer unused FAR to receiving sites subject to overall limits on the total increase in FAR allowed by TDR and other bonuses.

Trees – Portland uses TDR to motivate the preservation of trees that are at least 12 inches in diameter and certified as healthy by the City Forester or a certified arborist. In Portland's multi-dwelling residential zones, the tree owner must covenant that the trees will be preserved for at least 50 years. A tree under covenant must be determined to be dead, dying, or dangerous by the City Forester or a certified arborist before being removed and must be replaced within 12 months of removal. Table 120-4 spells out the amount of floor area that can be transferred for three categories of tree diameter in four different zoning districts. At the low end of this table, 1,000 square feet of floor area can be transferred for each tree between 12 and 19 inches in diameter in the RM1 zone. At the high end, 16,000 square feet can be transferred for each tree 36 inches or greater in the RM4 and RX zones. However, the total floor area transferred cannot exceed the unused floor area of the sending site. The receiving area provisions in the multi-dwelling residential zones are limited in the total increase in floor area achievable via TDR and all other allowed bonuses.

Central City Plan District – The Central City Plan District alone consumes 108 pages of Portland's zoning code. In 2007, Portland commissioned a study to explore whether the numerous options for gaining additional development potential were accomplishing the goals of this district (Johnson-Gardner 2007). The report characterized the Central City district as having 18 bonus options and six transfer options adopted over the span of 20 years. The bonus options at that time included additional floor area for projects that provide: 1) residential units in target areas; 2) locker rooms; 3) day care; 4) rooftop gardens; 5) percent for art; 6) water features/fountains; 7) eco-roof; 8) middle-income housing; 9) affordable housing; 10) Willamette River Greenway; 11) large household dwelling units; 12) open space; 13) open space funds; 14) large dwelling units; 15) small development site; 16) below grade parking; 17) retail space; and 18) theater space within Broadway Theater Target Area. The Johnson-Gardner study identified the following six mechanisms as Central City FAR Transfer Options as they existed in 2007.

1. *Abutting Lots Transfer* – This option allowed FAR potential to be shifted between abutting lots allowing additional density on one or more of these lots as a way of creating cohesion and place making in larger developments. Generally, maximum transfer was 3:1 FAR bonus.

2. *Single Room Occupancy (SRO) Transfer* – The unused FAR from a building covenanted to remain SRO could be transferred and used to generally gain an extra 3:1 FAR at receiving sites throughout the Central City subject to certain restrictions: FAR from RX could transfer to RX, EX or CX districts but could not receive FAR transferred from the EX or CX zones. The Single Room Occupancy (SRO) housing option was used at least once, successfully preserving the Athens Hotel as SRO units in 1990.

3. *Historic Landmarks Transfer* – Unused floor area from a designated landmark could be transferred to receiving sites in the same neighborhood but not further than two miles according to the 2007 study. The receiving site could gain up to a 3:1 FAR bonus with some exceptions.

4. *Residential Floor Area Transfer* – Owners of residential units in the Central City could transfer unused FAR allowing a general increase of FAR 3:1 on receiving sites anywhere in the Central City. The 2007 study noted that this mechanism was adopted at a time when Central City residential development had stalled. The Johnson-Gardner study considered this to be the most versatile transfer option given the large number of potential sending sites.

5. *South Waterfront Transfer* – This option allowed abutting or non-abutting receiving sites in the South Waterfront sub-district to achieve extra FAR and to exceed the typical bonus of 9:1 FAR maximum for FAR transferred from the Willamette River Greenway. The Johnson-Gardner study noted that the ability to exceed 9:1 FAR made this option particularly attractive and that it had experienced some use by 2007.

6. *Central City Master Plan Transfer* – This option allowed FAR from multiple sites to be distributed throughout a master planned development that could be several blocks in size. Unlike the Abutting Lot Transfer, this option could involve parcels throughout the Central City and there was no limit on the FAR that could be transferred and used on receiving lots as long as the city determined the proposal to implement the Central City Plan. For example, the Central City Master Plan tool was used to shift development potential from The Edge, a building housing an REI outlet, to a larger building known as the Elizabeth located seven blocks away. In another example, the five-block area known as the Brewery Blocks were treated as one site, allowing FAR to shift in a way that transferred FAR from two landmarks, the Brewhouse and Armory, to new buildings in this complex.

The 2007 report concluded that the six transfer options and 18 bonus options were in competition with each other and that developers logically preferred to use the cheapest way of gaining additional FAR. Comparing the 18 bonus options with one another, the Johnson-Gardner study found that the locker room bonus was particularly attractive to developers since it generated 40 square feet of bonus floor area per square foot of amenity, a higher ratio than all the other 17 options combined. On a cost per square foot of bonus floor area basis, the locker room option (at $6 per square foot of bonus floor area) and the eco roof option (at $8 per square foot) were cheaper than all the other bonus FAR options. With the exception of a bonus for market-rate residential, (which was not likely to change developers' plans), the report noted that the locker room and eco roof bonus options were used more frequently than other bonus FAR options. The locker room and eco roof FAR options were also cheaper than the transfer FAR options, which Johnson-Gardner estimated at ranging from $6.50 to $18.00 per square foot and averaging $10.00 per square foot in 2007.

This report also reconfirmed that developers gravitate toward bonuses that add value to their projects rather than community benefits at off-site locations. While some of the bonus FAR options may not recoup their cost, TDR options are even less likely to directly add value exclusively to a receiving site because the community benefit generated at a TDR sending site is often not close to the receiving site.

As of March 2021, Section 33.510.205 of the Central City Plan District prioritizes options by specifying which bonus and transfer options must be used before others. For example, the first tier of 3:1 FAR bonus must come from inclusionary housing, affordable housing, historic landmark transfers or riverfront open space. There are some exceptions to this prioritization requirement. For example, the South Waterfront subdistrict is exempt from this prioritization, but for certain projects in target areas, the South Waterfront Willamette River Greenway bonus option must be used before any other bonus.

After the first 3:1 FAR of bonus, projects can use other bonus floor area options: the inclusionary housing bonus option, the Affordable Housing Fund bonus option, the riverfront open space bonus option, the South Waterfront Willamette River Greenway bonus option, the South Waterfront open space bonus option, and payment to the SWPOSF. Portland distinguishes the bonus options from its transfer options even though the public benefit from a bonus option might occur offsite, an outcome that in many other jurisdictions would be characterized as a form of transfer rather than a bonus, which typically means allowing added development potential to a project in return for a public benefit provided at that project site.

Importantly, the code as it existed in March 2021 listed six FAR bonus options rather than the 18 options available in 2007. This suggests that Portland responded to the Johnson-Gardner study by limiting bonus FAR to achieve community benefits that were most important and/or in the greatest need of incentivizing. As detailed below, the code in March 2021 requires certain developments to provide locker rooms and eco roofs without getting a FAR bonus in return. In addition to requiring rather than incentivizing locker rooms and eco roofs, this change eliminates the two ways of gaining bonus FAR that were cheaper than the average cost of transferring FAR.

As of March 2021, the Central City transfer options at 33.510.205.D do not limit the amount of floor area that can be transferred to a receiving site as long as the sending site retains the minimum FAR required by the code or an amount equal to the total surface parking area multiplied by the maximum allowed floor area, whichever is more. Two transfer options are discussed in this code section.

Historic Resource Transfer Option – Potential sending sites consist of Landmarks or contributing resources in a historic or conservation district that are zoned RM3, RM4, RX, CX, EX or OS. Sending site owners who choose to participate must enter into an agreement to bring the structure up to seismic standards if it is not already compliant and enter into a covenant prohibiting the demolition or removal of the structure unless approved by the city. Qualified sending sites can transfer their unused FAR plus an extra 3:1 FAR. The receiving sites must be zoned RM3, RM4, RX, CX or EX and located within the Central City but outside the South Waterfront subdistrict. However, there are exceptions that allow transfers of FAR outside the Central City if certain standards are met.

Transfer of Floor Area Within a Floor Area Transfer Sector – Floor area in the RX, CX, EX, and OS zones, including bonus floor area, can be transferred to any receiving site within the same floor area transfer sector as established by Map 510-23. The sending site cannot be a landmark or a contributing resource in a historic or conservation district. If bonus floor area is included in the transfer, the community benefit must be completed before the receiving site can be issued an occupancy permit.

As of March 2021, Section 33.510.243 mandated all buildings of 20,000 square feet or more in three zones to provide eco roofs as a requirement rather than as a way to achieve bonus density. Similarly, as the code existed in March 2021, in the South Waterfront sub-district, additions of at least 100,000 square feet of non-residential floor area had to, with some exceptions, provide locker rooms with showers, dressing areas, and lockers available to all building tenants without getting bonus FAR in return.

Reference

Johnson-Gardner. 2007. Evaluation of Entitlement Bonus and Transfer Programs: Portland's Central City. Accessed 2-27-21 at Microsoft Word - Bonus Study Report draft (oregonlive.com).

Providence, Rhode Island, population 179,883 (2019), uses TDR to preserve downtown historic landmarks. To participate, a restriction must preserve the landmark in perpetuity unless the receiving site project is demolished, in which case the transferred height can be reclaimed by the sending site. The difference between the height of the preserved landmark building and the height permitted by zoning can be used to exceed baseline height on a receiving site by 1.6 times or by 300 feet, whichever is less.

Puget Sound Region, Washington, includes four counties surrounding Seattle in the northern part of the state: King, Pierce, Kitsap, and Snohomish. Each of these counties has its own individual TDR program with the King County program being the oldest of the four and preserving over 144,500 acres as of March 2021, making it the most successful TDR program in the nation in terms of amount of land preserved.

The regional program was initiated by key legislators and Forterra, (formerly the Cascade Land Conservancy), a regional conservancy working in ten western counties in the State of Washington. In 2007, the legislature passed a measure to study the feasibility of a regional TDR market. In 2008, the Washington office of Community Trade and Economic Development (CTED) completed a report recommending that the state proceed with a TDR program for the Central Puget Sound region within specified parameters. In May 2009, Governor Chris Gregoire signed legislation that established this program.

The program is designed to encourage the voluntary transfer of development rights from rural, forested, and agricultural lands under county jurisdiction to receiving sites in cities and towns throughout the four-county region. Individual cities/towns and counties may choose to use a preexisting code section to establish interjurisdictional TDR programs and enter into separate interlocal agreements as they have in the past. Alternatively, communities may choose to use the components for regional agreements developed by CTED and codified in WAC 365-198, which became effective in 2010.

In addition to the benefits generated by all TDR programs, participation in the Puget Sound program allows cities with populations of at least 22,500 in King, Pierce, and Snohomish counties to use the Landscape Conservation and Local Infrastructure Program (LCLIP), adopted in 2011. LCLIP allows eligible cities that meet regional TDR standards to fund infrastructure in qualified receiving areas that accommodate interjurisdictional transfers from counties in the region using tax increment financing, a tool not otherwise available in the State of Washington.

These three counties calculated the number of TDRs on agricultural and forest land of long-term commercial significance. The Puget Sound Regional Council (PSRC) allocated these TDRs to cities eligible to receive TDRs based on growth projections and other criteria. To qualify for LCLIP, a participating city must agree to accept at least 20 percent of its TDR allocation, adopt an infrastructure plan for the receiving area capable of accommodating these TDRs, and establish at least one infrastructure project area. Cities are able to receive a greater proportion of tax increment revenues by agreeing to accept more than 20 percent of the allocation.

King County has adopted TDR interlocal agreements with the cities of Seattle, Issaquah, Bellevue, Normandy Park, and Sammamish. Pierce County has adopted a TDR interlocal agreement with the City of Tacoma. Snohomish County has adopted a TDR interlocal agreement with the City of Mountlake Terrace.

The 2013 TDR Agreement between King County and Seattle won a Lifetime of GMA (Growth Management Act) Award in 2015. This agreement establishes a process in which development in Seattle's downtown and South Lake Union district can buy up to 800 urban density credits worth $18 million representing the preservation of 2,000 acres of farmland and 23,000 acres of forestland under King County jurisdiction. In return for Seattle's acceptance of these credits, King County will partner in the funding of infrastructure improvements within the city's receiving areas. For up to 25 years, the county will share 17.4 percent of the new property tax revenues generated in these two receiving areas. This shared tax revenue will help pay for roughly $17 million in Green Street improvements including transit, bicycle, and pedestrian facilities. King County also estimates that the redirection of 800 rural homes to Seattle's South Lake Union district will reduce greenhouse gas emissions by 173,000 metric tons as well as reduce the need in rural King County to build and maintain roads, sewers, and water mains as well as provide emergency and other public services (King County 2013).

Reference

King County. 2013. County Executive proposes strengthening "farm-to-city" connection with Seattle. News Release dated July 8, 2013. Accessed 3-21-21 at County Executive proposes strengthening "farm-to-city connection" with Seattle - King County.

Queen Anne's County, Maryland, population 50,381 (2019), lies on the east side of the Chesapeake Bay approximately 70 miles east of Washington, D.C. and 70 miles southwest of Wilmington, Delaware. The county is mainly composed of farms and wooded areas dotted with small towns and exurban residences.

In 1987, Queen Anne's County adopted a new comprehensive plan implemented by zoning that limited rural land to an on-site density of one unit per ten acres or one unit per eight acres if development was clustered on 15 percent of the property. The code also introduced a TDR program allowing TDRs to be transferred from sending sites zoned AG or Countryside (CS) to receiving sites zoned AG, CS or Suburban Estate (SE), thereby allowing these sites to achieve a maximum density of one unit per nine acres. Between 1987 and 1994, brisk activity in this program was largely attributed to the fact that properties in the AG district could become receiving as well as sending sites.

Another program known as Non-Contiguous Development (NCD) was adopted in 1987. It allowed non-contiguous properties in the AG and CS zones to be treated as a single development for the purpose of transferring density between sending and receiving sites in the same zoning district. The NCD mechanism allowed receiving site densities as high as one unit per acre and motivated the cross-county transfer of TDRs from the relatively remote northeastern corner of Queen Anne's County to receiving areas near the Chesapeake Bay Bridge where growth pressure is higher due to access to Annapolis and Washington, D.C. (McConnell, Walls & Kelly 2007).

In 1994, the receiving area for the TDR program was limited to designated Growth Areas, a change that reduced transfers under that program because the potential receiving area shrunk from 209,000 acres to 6,400 acres and because the baseline densities in the Growth Areas were higher than market demand. Even though the 1994 amendments to the TDR program created potential commercial receiving sites, developers reportedly found the intensity bonus with TDR too small to spend the time and effort needed to navigate the TDR process (McConnell, Walls & Kelly 2007).

In 2004, Queen Anne's County added Critical Areas, an overlay zoning district with an on-site density limit of one unit per 20 acres, as sending areas in the TDR program. The receiving areas could also be located in the Critical Area overlay, with individual densities reaching one unit per five acres as long as the overall density of the Resource Conservation Area did not exceed one unit per 20 acres. Critical Area TDRs were in high demand because they could boost density in the desirable waterfront properties zoned Critical Area Overlay. The value of Critical Area TDRs started at $35,000 each but rose as high as $265,000 each by 2005 as supply dwindled. By 2005, the TDR and NCD programs had protected almost 10,000 acres. A 2007 study concluded that Queen Anne's County's programs demonstrate that rural receiving areas can create substantial demand for the additional density provided by TDR (McConnell, Walls & Kelly 2007).

As of January 2021, the county's TDR code, Article XX Section 18:1-100 – 107, states that sending areas outside the Chesapeake Bay Critical Area must be at least 24 acres in size and meet specific criteria for soil and/or woodland classifications. Within the Chesapeake Bay Critical Area, sending parcels must be at least 20 acres in size. The Planning Director is tasked with reviewing applications and issuing TDR certificates. Transfers are approved administratively. Sending sites can generate one TDR per eight acres of preserved land in the AG and one TDR per five acres of preserved land in the CS district.

The code now allows receiving areas in over a dozen zoning districts and specifies how much receiving site bonus development can be achieved depending on the zoning classification of the sending site and whether or not the sending and receiving sites are outside or within the Critical Area Resource Conservation Area. For example, receiving sites outside the critical area can be approved by cluster or planned residential development in the E, SE, SR, UR, VC, GNC, SHVC, GVC, TC, and CS zones. In these zones, transferred TDRs allow receiving site open space to be decreased up to 25 percent or density and net building area increased up to 25 percent. On the other hand, on receiving sites zoned E, SE, SR, UR VC, GNC, SHVC, GVC, and TC within the critical area, transferred TDRs can reduce open space up to 25 percent or increase density and net buildable area up to 25 percent as long as 20 acres of critical area RCA are preserved per TDR.

TDRs can also increase non-residential floor area. For example, on non-residential receiving sites in a critical area zoned VC, TC, SC, UC, and SI, TDR allows baseline floor area and impervious surface area coverage to be increased up to 25 percent. Different ratios apply when sending areas are zoned AG, zoned CS within the critical area, or zoned CS and located outside the critical area.

As of 2016, Queen Anne's County's TDR programs had preserved 28,230 acres. Some authors attribute this program's early activity to the fact that receiving sites could be located in the AG zone in the first few years of the program. But even after the county removed the AG zone as a potential receiving area, the program continued to be reasonably successful despite difficulties in transferring density to Kent Island, seemingly the most logical receiving area due to its location at the end of the Chesapeake Bay Bridge.

Queen Anne's County motivates sending area property owners to transfer density with the transfer ratio created when a sending property's total acreage can be included in the TDR allocation formula even though some of that land might be undevelopable due to poor soil suitability for septic systems. Administrative approval of transfers similarly motivates receiving site developers who might otherwise be concerned about the uncertainty, delay, and cost often resulting from public hearings and discretionary processes. The county also generates preservation by allowing additional development in desirable locations in return for a considerable amount of sending site preservation.

References

McConnell, V., M. Walls, and F. Kelly. 2007. Making TDR Programs Better: Report Prepared for the Maryland Center for Agroecology. Accessed 1-29-21 at Microsoft Word - McConnell Walls.FINAL-2.doc (umd.edu).

Maryland. 2016. Transfer of Development Rights Committee Report. Maryland Department of Planning. Accessed 1-29-21 at TDR-committee-report-2016.pdf (maryland.gov).

Raynham, Massachusetts, population 14,313 (2018), uses TDR to preserve open space, farmland, natural areas, recreation, and water supply. TDRs can be used to exceed receiving site residential densities from 50 to 100 percent.

Red Hook (Town), Dutchess County, New York, population 11,181 (2016), has open space incentive zoning allowing increased density in the Traditional Neighborhood Development District in return for preserving farmland and environmentally sensitive areas in the Agricultural Business District. Residential receiving sites add one unit per three acres preserved or a DTC of $20,000. Non-residential adds 2,000 square feet of floor area per two preserved acres or $15,000 DTC. The town automatically adjusts DTCs annually based on changes in median home sales price.

Redmond, Washington, population 71,929 (2019), lies 20 miles east of downtown Seattle. In 1995, the City adopted a TDR program designed to preserve agriculture and recreational land in the Sammamish River Valley. In the next 12 years, the TDR program was expanded to accomplish additional goals including the protection of environmentally sensitive areas including streams, and wetlands, as well as historic sites. In addition to having a variety of sending areas, the Redmond TDR program by 2007 offered developers a variety of urban receiving areas plus a wide range of incentives to use TDR including increased floor area and building height as well as modifications of standards for open space, parking, and impervious surface coverage.

As of January 2021, the Redmond zoning code (20D.200.10) contained four TDR sending area categories: 1) Property zoned Urban Recreation; 2) Historic sites; 3) Environmentally critical areas (specifically Species Protection Areas, selected wetlands, stream and stream buffers, and landslide hazard areas plus buffers); and 4) Forested land meeting criteria for area, native species, and tree age.

The allocation of TDRs varies depending on the zoning designation of the sending sites. At the low end, 1.1 TDRs can be generated per acre of sending area land zoned Urban Recreation. At the high end, 13.8 TDRs per acre can be produced on sending sites in any Downtown district or 14 TDRs per acre of land zoned Overlake Village or Overlake Business and Advanced Technology.

Receiving areas are properties in the following zones: all Downtown districts, Overlake Village, General Commercial, Overlake Business and Advanced Technology, Gateway Design, Business Park, Manufacturing Park, and Industry.

In receiving site projects, one TDR can accomplish any of the following: 8,712 square feet of additional floor area; substitution for a requirement to provide 8,712 square feet of park land; 8,712 square feet of additional lot coverage or impervious surface coverage (with the increase not greater than ten percent of the site); 8,712 square feet of additional floor area in a story exceeding baseline building height (with restrictions in some zones). In addition, one TDR can allow receiving site developers to add up to five stalls above Redmond's maximum parking standards not to exceed a total of five stalls per 1,000 square feet of floor area.

As of 2008, the Redmond TDR program had transferred 573 TDRs worth over $16.8 million, protecting more than 420 acres of farmland, urban open space, and environmentally sensitive sites (Churchill 2008). As of April 30, 2018, 961 TDRs had been transacted with a sale total over $24 million, yielding an average mean of $26,396 per TDR. Microsoft Corporation was the largest customer, accounting for over 805 TDRs purchased. Sales prices have varied, ranging from a low of $9,246 per TDR in a 2010 transaction to a high of $70,000 per TDR for a fractional TDR transaction in 1997.

References

Churchill, Jeff. 2008. Redmond's TDR Program Aims to Preserve Rural Sammamish Valley. Accessed 2-26-21 at About-Growth---Fall-2008-PDF (redmond.gov).

Rhinebeck, New York, population 2,563 (2019) uses TDR to preserve farmland, open space, environmental areas and scenic values recognized by the Hudson River National Historic Landmark district and other scenic designations.

Rice County, Minnesota, population 66,972 (2019), is a largely agricultural jurisdiction located 30 miles south of Minneapolis. In 2004, the county adopted a TDR program aimed at preserving farmland, environmentally sensitive areas, and open space. By 2016, the program had succeeded in preserving 5,862 acres of land largely because of the strict development limitations that apply here unless property owners choose to use the TDR option.

The Agricultural zone, Rice County's largest district, allows one dwelling per quarter-quarter section. Under some circumstances, a second unit can be permitted per quarter-quarter section. But the fact remains that many property owners are motivated to forego this limited development potential in quarter quarters with productive soils, particularly on quarter quarters that are within the interior of farms rather than on public roadways, if they can sell that development potential in the form of TDRs.

Land zoned Agricultural District can become sending sites, generating one TDR for each foregone unit allowed by zoning (with several exceptions, such as additional TDRs for undeveloped parcels of record of various sizes). When these TDRs are used in smaller receiving site projects, this represents a one-to-one transfer ratio. However, only four TDRs are needed to produce a receiving site project involving five bonus units, which creates a 1.25-to-1 transfer ratio that several receiving site developments here have employed.

In addition to the Agricultural District, the Rice County TDR program allows sending sites in four other zones. In the Urban Reserve District, minimum lot area is 35 acres. Minimum riparian lot area is 20,000 square feet in the General Development Shoreland (GDS), 40,000 square feet in the Recreational Development Shoreland (RDS), and 80,000 square feet in the Natural Environment Shoreland (NES). As in the Agricultural district, TDR allocation is the number of units allowed by the sending site's zoning.

After TDRs have transferred, the sending site is restricted from further development. However, if the site is rezoned to a less restrictive zone, development potential is determined by the new zoning minus the number of units transferred. If a sending site is annexed to a city, the development restriction associated with a TDR transfer is removed.

TDRs from sending sites in the Agricultural district can be used in three types of receiving site projects on land that is also zoned Agricultural projects: minor cluster developments, golf course cluster developments, and planned unit developments being built as village extension areas on land with an Agricultural zoning designation that will be (or has been) changed to the Village Mixed Use District.

TDRs from Shoreland zoning districts can only be transferred to a golf course cluster development, a village extension area within the Agricultural district, or to a receiving site in the same shoreland district around the same lake.

Sending and receiving sites must be located within the same township except when these sites are contiguous, under common ownership, and when the transfer is authorized by both townships.

As of 2016, all 14 townships in Rice County experienced some TDR activity in varying degrees ranging from a high of 32 receiving lots created in one township versus a low of two receiving lots in another township. In 2016, a total of 192 receiving lots had been created from a total of 175 sending lots. The 17 additional receiving site lots occurred because receiving projects with five bonus lots only had to transfer four TDRs. The total sending area protected by TDR was 5,862 as of 2016. That represents over 33 acres preserved per TDR, demonstrating the benefit of strong development controls in sending areas where the community is serious about protecting significant resources such as farmland.

Ridgeland, South Carolina, population 3,911 (2019), retained the TDR mechanism in its Smart Code as a way of preserving flood plains, steep slopes, woodlands, farmland, view sheds, open space, corridors and buffers. In four receiving area zones, TDR can be used to increase baseline housing unit densities by from 12.5 to 100 percent. Each housing unit can be converted to two bedrooms or 1,000 square feet of retail/office floor area.

Riverhead, New York, population 33,539 (2018) uses TDR to preserve farmland. TDRs can increase density on receiving sites in four residential zoning districts. In four commercial zones, a mixed-use zone and the Planned Recreational Park zone, one TDR yields 1,500 square feet of added floor area. As of 2014, about 121 acres had been preserved. Riverhead also sends and receives Pine

Barrens Credits in the regional Central Pine Barrens TDR program.

Robbinsville Township, Mercer County, New Jersey, population 14,543 (2019), uses non-contiguous planned unit developments to preserve priority open space and transfer development credits to its Town Center. Proceeds from sales of credits from open space owned by the township can fund future preservation.

Sammamish, Washington, population 65,733 (2018), uses TDR to achieve various community benefits in three programs. In the in-city program, TDRs are transferred from sending sites with important resources, erosion hazards and wetlands to receiving sites in the city's Town Center. The Town Center can also accept 75 TDRs from sending sites known as the "emerald necklace" that are under King County jurisdiction according to provisions of an interlocal agreement between the city and county. In a third program, TDRs can be transferred from a sending zone to a receiving zone that are both in Town Center.

San Antonio, Texas, population 1,487,000 (2018), has retained the TDR mechanism in its Smart Code as a way of preserving flood plains, steep slopes, woodlands, farmland, viewsheds, open space, corridors and buffers. In four receiving area zones, TDR and/or sustainability design can be used to increase the baseline housing unit density. When the bonus option is used, the receiving site development must at least double the baseline unit density and maximum densities are not prescribed. One housing unit converts to three bedrooms of lodging or from 750 to 1,500 square feet of office/retail floor area.

San Bernardino County, California, population 2,180,000 (2019), uses the planned development process (84.18.030) to transfer density that preserves valuable resources or provides public amenities beyond normal expectations.

San Diego, California, population 1,423,852 (2019), has a TDR program that promotes the creation of parks and the preservation/restoration of designated historic resources in the Centre City Planned District. The City Manager has sole discretion to approve transfers to receiving sites or a TDR bank. Developers can also gain FAR bonuses by providing affordable housing, urban open space, eco-roofs, three-bedroom residential units, employment uses, green building design, public parking, and a FAR payment bonus that allowed extra floor area at $18.67 per square foot in 2018-19. In its first five years, the payment option had generated $1.7 million for public parks and enhanced right-of-way improvements.

San Francisco, California, population 874,671 (2019) arguably has the most successful historic preservation TDR programs in the nation. The full profile can be found in Chapter 5, Historic Landmarks, in Part I.

San Luis Obispo County, California, population 283,111 (2019), lies 230 miles south of San Francisco and 190 miles north of Los Angeles on the Pacific Coast.

The county's first TDC program is managed by a private non-profit conservancy, which by 2016 had purchased over 350 lots in the coastal community of Cambria using a small loan to launch an ongoing revolving fund for preservation (Johnson 2016). The Cambria TDR program appears in Chapter 8, Biodiversity, in Part I. A second program, profiled here, has preserved 5,464 acres in the inland portions of the county using a TDC allocation process involving site-specific appraisals in order to adjust for the wide range of easement values encountered in a large and diverse sending area.

Studies conducted in the 1990s found 23,000 undeveloped lots within unincorporated San Luis Obispo County, with 12,000 of these lots in rural areas. Many of these were substandard lots within antiquated subdivisions. At that time, the county's general plan permitted an additional 8,000 rural lots. Severe adverse impacts were predicted if this development potential was actually used. As one of many responses, the county adopted its countywide TDC program in 1996 to reduce the inadvisable lots in antiquated subdivisions as well as protect environmentally sensitive areas and agricultural land by transferring development potential to places more suitable for growth.

Sending sites are proposed by the property owners and reviewed for eligibility by the Planning Commission based on whether they satisfy at least one specific criteria or general criteria in three areas.

Agricultural Criteria
- Specific:
  - Land capability: sending site at least 40 acres in size and 50 percent Class I or II soils
  - Grazing land: minimum 320 acres with at least 100 acres well or moderately suitable for rangeland
- General: Land with
  - Demonstrated productivity
  - Microclimates suitable for specific crops
  - Localized groundwater dependency
  - Soil conservation benefits.

Natural Resource Criteria
- Specific:
  - Designated by county plans as Natural Area or Significant Habitat
  - Open space adjacent to restricted open space
  - Protects views from highways and main collector streets
- General: Land that
  - Protects watersheds or conserves soil
  - Protects proposed greenbelts, community separators, scenic entries, or natural resources identified by the county and/or local communities
- Provides access to natural features for hiking, nature education, or other passive recreation

Antiquated Subdivision Criteria
- Specific:
  - Sites at least ten miles from urban or village reserve lines with lots smaller than 20 acres
  - Sites at least five miles from urban or village reserve lines with lots smaller than ten acres
  - Sites located within designated antiquated subdivisions
- General:
  - Sites within antiquated subdivisions with substandard improvements
  - Sites that are distant from transportation and other services and therefore conflict

with air quality goals and increase the cost of public services

TDC Allocation

Sending area applicants can choose between two methods of determining the number of TDCs available to transfer.

Existing Lots Method – One TDC for each legal lot proposed for retirement; no more than one TDC can be granted for retirement of a substandard lot that meets a specific natural resource criterion

Development Value Method – The appraised development value of the sending site is divided by twenty.

- Development value is the full value of the sending site if a public agency or non-profit organization agrees to accept title to the property
- If the owner proposes to retain title, development value is full value minus the value after restrictions imposed by proposed easements

Following the recordation of necessary easements, the TDC Administrator issues a Certificate of Sending Credits documenting the number of TDCs and a unique registration number for each credit. Upon each certificate transaction, the TDC Administrator makes the needed certificate revisions. Holders of TDCs sell them at prices negotiated between buyers and sellers; the formula in which appraised sending site development value is divided by $20,000 only determines the number of TDCs allocated to a sending site.

Receiving Sites

Receiving sites are proposed by applicants and are only eligible if they meet all of the following eight criteria.

- Environmental review indicates no significant, unavoidable adverse impacts
- Not in an Agricultural land use category
- Within an urban or village reserve line or a Community-Based TDC area
- Less than 30 percent slope
- Outside Sensitive Resource Areas and hazard areas for flooding, geology, earthquakes, or wildfires
- Outside natural or habitat areas defined by the Land Use Element
- Project complies with all development and land division standards
- The site was not an approved sending site

One TDC is required for each additional dwelling unit allowed by a general plan amendment or land division through a parcel map or tract map. However, the code allows the Review Authority to waive TDC requirements for general plan amendments consistent with strategic growth policies, projects with affordable housing, or sites with other special circumstances.

Base density is the amount of development potential allowed to the receiving site for minimum parcel sizes established by County Code. For receiving sites that are outside city limits but within an incorporated city's urban or village reserve line, the receiving site must be supported by the city and the maximum density allowed via TDC must be consistent with that city's policies, programs and standards but cannot exceed 50 percent of base density. However, if the receiving site project proposes amenities that go beyond basic requirements, an additional 25 percent of base density bonus may be granted.

San Luis Obispo aims to have receiving sites as close as possible to sending sites. If they are available, TDCs must come from sending sites no more than five miles from their receiving sites. When insufficient credits are available within five miles, the sending and receiving sites can be from any planning area in the same geographical region. However, in the case of the South County (Inland) and Huasna planning areas, the sending and receiving sites must be in the same planning area.

As of 2007, Chapter 22.24.300 created a community-based program for South Atascadero that mainly uses the county-wide procedure but involves special requirements for receiving areas regarding minimum site size, percolation tests, groundwater monitoring, community water, setbacks, drainage, and tree removal, as well as surveys for botanical and archeological resources.

Performance

As of 2012, the countywide program had protected a total of 5,464 acres (Nelson, Pruetz & Woodruff 2012). The two largest sending sites are the Bonheim Ranch, which alone conserved 5,364 acres and the Black Lake Canyon Preserve owned by the Land Conservancy of San Luis Obispo County. In 2018, the Land Conservancy reported that the market could support a value of $25,000 per credit (San Luis Obispo County 2018).

References

Johnson, Jay. 2016. Request by the Department of Planning and Building for authorization to process updates to the Coastal Zone Land Use Ordinance and the North Coast Area Plan to expand the Cambria Transfer of Development Credits Program. Memo dated 11/15.2016 from Jay Johnson, Planning and Building to Board of Supervisors.

Nelson, C., R. Pruetz, and D. Woodruff. 2012. The TDR Handbook: Designing and Implementing Transfer of Development Rights Programs. Washington, D.C.: Island Press.

San Luis Obispo County. 2018. Information Memorandum – Inland Transfer of Development Credit Program. Memo from Department of Building and Planning to Board of Supervisors dated February 20, 2018.

San Mateo County, California, population 767,423 (2019), allows owners of prime agricultural land in its Planned Agricultural District to transfer density plus bonus credits generated by lot mergers and agricultural water improvements.

Santa Barbara, California, population 91,376 (2019) encourages property owners to demolish existing oversized buildings, replace them with new, code-compliant buildings, and transfer the foregone floor area to receiving sites not for the purpose of building extra intensity but to be able to avoid the city's annual growth limits. Transfers to at least five receiving sites have occurred.

Santa Fe County, New Mexico, population 150,398 (2019), replaced a 2001 partial-county program with a countywide TDR program in 2016 aimed at preserving agriculture, historic resources, environmental areas, open space, and scenic vistas. One TDR is issued for retaining valid irrigation water rights on one acre of land as well as for eliminating the potential to build one dwelling unit on a sending site. Each TDR allows receiving site projects four bonus dwelling units, 10,000 square feet of additional non-residential floor area, plus various relaxations of lot coverage, building height, lot width, and the percent of residential development in seven categories of receiving areas including land rezoned to a higher density. As of 2020, ten TDRs had been certified.

Santa Monica, California, population 91,577 (2019), allows projects in its Bergamot Area to use higher tiers of FAR and other development standards by providing parks, art, and other community benefits off-site as well as on-site.

Sarasota County, Florida, population 433,742 (2019), lies 50 miles south of Tampa on the Gulf Coast of Florida and uses TDR to reduce vulnerability to the increased risk of storm surge caused by sea level rise and climate change as well as implementing many other goals including the preservation of environmentally-sensitive areas, agriculture, parcels of historic of archeological significance, and barrier islands as well as reducing the large number of substandard lots in antiquated subdivisions. The complete profile for Sarasota County can be found in Chapter 6 on sea level rise in Part I.

Scarborough, Maine, population 20,352 (2018), uses TDR to preserve natural areas, open space, farmland, forests and potential recreational areas. Receiving area developments can use a traditional transfer mechanism or pay a development transfer fee, set at $20,000 per dwelling unit as of 2020, that the town uses to preserve qualified sites by fee acquisition or permanent easement. The fee must be applied proportionately to every unit in a receiving site and paid prior to issuance of a building permit for each dwelling unit or residential lot.

Scottsdale, Arizona, population 250,602 (2019), uses density transfer (Section 6.1081) to shift development potential from severely- to less-constrained non-contiguous parcels within its Environmentally Sensitive Land (ESL) district.

Seattle, Washington, population 724,305 (2019), is the largest city in King County. In 1985, the city launched its first TDR program in its downtown, aimed at promoting affordable housing, open space, historic preservation, performing arts centers, and landmark theaters that include housing. In addition to downtown, Seattle subsequently expanded its use of TDR into most of its South Lake Union district and parts of its University, Uptown, and North Rainier neighborhoods. Seattle's TDR bank has bought and sold floor area representing the preservation of affordable housing, landmarks, and landmark performing arts centers as well as the development of its concert hall and sculpture park. Seattle has cooperated with King County on interjurisdictional transfers and participates in a three-county regional TDR program that allows the city to use an infrastructure financing tool that would not otherwise be available.

Downtown TDR

Seattle's 1985 downtown plan reduced base FAR but allowed developers to gain bonus floor area by transferring it to receiving site projects to accomplish various planning goals. As of 2021, TDR regulations had been amended at least 14 times and Section 23.49.014 of Seattle's Land Use Code offered seven different types of TDR:

- Within-Block TDR allows transfers between any lot in the same block.
- Housing TDR

- DMC (Downtown Mixed Commercial) Housing TDR
- Landmark Housing TDR
- Landmark TDR: Eligible sites are designated landmarks pursuant to the city's Landmark Preservation Ordinance.
- Open Space TDR: Open space that is designed, maintained, and operated similar to a public park
- South Downtown Historic TDR: Eligible sending sites are structures contributing to the architectural or historic character of the Pioneer Square Preservation District or the International Special Review District.

Table 23.49.014A of the Seattle Municipal Code displays six forms of TDR operating within some or all of twelve zoning categories, producing 72 different permutations. For example, lots in the DMC 170 district can serve as sending or receiving sites for the Housing TDR, DMC Housing TDR, Landmark TDR/Landmark Housing TDR, and Open Space TDR options but can only serve as receiving sites for the South Downtown TDR program.

With the exception of South Downtown Historic, Landmark Housing and Landmark TDR sending sites, transferable floor area is calculated by subtracting the existing floor area from the amount of floor area allowed to the sending site by the base FAR. For Landmark Housing TDR sending sites, transferable floor area is the total floor area permitted by code without deducting existing floor area. However, if the landmark exceeds the base FAR, transferable floor area is the floor area allowed by code minus the floor area that exceeds base FAR. In all cases, floor area previously transferred is deducted from the remaining transferable floor area.

In South Downtown, sending sites in two zones can transfer six times the floor area minus existing floor area. In all other South Downtown zones, the transferable floor area is the floor area allowed by base FAR minus existing floor area. In addition, South Downtown sending sites cannot transfer more than three times their lot area. Several additional regulations and exceptions are in the code.

Transfers from Landmark and Landmark Housing TDR sending sites require the historic building to be rehabilitated and maintained as directed by the Landmarks Preservation Board or, in the case of South Downtown Historic TDR, as required by the Director of Neighborhoods upon recommendation from the Special Review District Board or the Pioneer Square Preservation Board. In the Housing TDR program, sending sites are required to be rehabilitated with the goal of assuring at least an additional 50 years of useful building life. If Housing or Landmark Housing TDRs are proposed to be transferred prior to completion of rehabilitation, a security may be required in order to assure completion. The owner of a Housing, Landmark Housing, or DMC Housing TDR sending site must enter into an agreement to provide maintenance and adhere to rent and occupancy requirements for at least 50 years. The transfer of TDRs from a sending area remains in effect for the life of the project at the site that received these TDRs.

Code Section 23.58A.042 includes receiving site requirements that vary depending on whether the transfer preserves landmarks, provides housing, adds open space or produces some other community benefit. For example, open space sending sites must meet 13 standards including design elements, public access, minimum hours of availability to the public, and solar exposure.

Receiving sites can be located in any of 11 zoning district categories. For example, receiving sites for Within-Block TDR, Housing TDR, Landmark TDR, Landmark Housing TDR, Open Space TDR and South Downtown Historic TDR can be approved in the Downtown Office Core 1 (DOC1),

DOC2, Downtown Retail Core (DRC), and Downtown Mixed Commercial (DMC) 340/290-440 zones. Base FAR and Maximum FAR are 6 and 21 in DOC1, 5 and 15 in DOC2, 3 and 6 in DRC, and 5 and 11 in the DMC 340/290-440 zone.

According to Table B for Code Section 23.49.011, the first increment of FAR above base must be achieved by acquiring regional TDR credits if the receiving site is within the Local Infrastructure Project Area. The amount of this first increment ranges from FAR 0.25 in the DMC zones up to 1.0 FAR in the DOC 1 zones. In DOC1, DOC2 and DMC zones outside the South Downtown, once a receiving site applies this first increment of bonus FAR, if applicable, the remaining allowable bonus FAR can be gained by providing combinations of TDR, housing, childcare, or project amenities as long as at least five percent of all bonus floor area (in some districts) comes from Landmark TDR, if available from the TDR Bank or private sources. In some zones and under some circumstances, a minimum of 12.5 percent of bonus FAR must be gained by TDRs from a major performing arts facility (if available from the city at or below TDR market prices) and another 12.5 percent from other forms of TDR except Housing TDR. Subject to many conditions and exceptions, Seattle exempts the floor area occupied by many uses and architectural features from FAR maximums including museums, performing arts theaters, showers for bike commuters, community centers, and schools. This level of complexity indicates that Seattle recognizes the tendency for developers to use the cheapest way of gaining bonus FAR unless the code requires the use of specified methods including TDR. It should also be noted that Section 23.49.011, (which pertains to only a portion of Seattle's receiving areas), was amended by 24 ordinances between 2001 and 2020, suggesting that the city is routinely fine tuning these requirements to get the desired results from its bonus options.

Regional Development Credits Program

The regional TDR program involving sending and receiving sites in three Puget Sound counties is detailed in the King County profile. However, it is noted here that Seattle Code Section 23.58A.044 includes two tables that control the amount of bonus residential and non-residential floor area that one credit can generate in Seattle depending on the origin of the credit. For example, 1,640 square feet of bonus residential square feet can be generated by one Agricultural credit from a sending site in King County, versus 420 bonus square feet per Agricultural credit from Pierce County, or 860 bonus square feet per Forest credit from Snohomish County (with the proceeds from forest credit sales used to buy new agricultural credits). These exchange ratios change after the first 200 credits are extinguished. By meeting the criteria for participation in the regional TDR program, Seattle became eligible to use a form of tax increment financing to fund infrastructure, a tool that is only available in the State of Washington to communities that satisfy the regional TDR program requirements. In 2016, the tax increment generated by the transfer of region TDR was projected to fund $30 million of infrastructure improvements, including Seattle's Green Streets Program, an initiative to rebuild rights of way in ways that prioritize active transportation and public space (Kinney 2016).

TDR Bank

Per Code Section 3.20.320, TDRs either purchased by the city or held for potential sale from city sending sites are considered to be in the city's TDR Bank. TDRs from Landmark, Housing, and Open Space TDR sending sites are eligible to be purchased by the TDR Bank, which is managed by the Department of Housing. The Housing Director sells TDRs at prices based on negotiation.

\Performance

As of 2010, 1,970,349 million square feet of floor area had been certified in the following five categories: Landmark Performing Arts Theater/Housing 241,144; Open Space 272,025; Landmark 301,496; Housing 648,648; and Major Performing Arts Theater 507,036. At that time, TDR had preserved ten landmarks including two landmark performing arts theaters, the Paramount Theater, and the Eagles Auditorium. By 2010, 19 buildings with affordable housing had been preserved by TDR (often in combination with other funding sources) with a total of 1,106 units (Meier 2010). As of 2018, another 316,653 square feet of floor area had been transferred into downtown Seattle from sending sites within unincorporated King County using interjurisdictional programs (Fesler 2018).

The TDR Bank has been essential to the success of TDR in Seattle. The Bank was the only buyer of TDRs in the program's first 12 years (Massachusetts Undated). The Bank's first sale was in 1996, when it sold 130,000 square feet to the W Hotel for $1.47 million. By 1998, the TDR Bank had purchased 274,340 square feet of floor area from eight buildings, including the Paramount Theater and the Eagles Auditorium, and had sold 249,380 square feet. In addition, the city had 423,000 square feet of Major Performing Arts Facility TDRs to sell, with the sale proceeds to be used to pay the debt service on construction of Benaroya Symphony Hall (Walker 1998).

In a single transfer to the Washington Mutual office tower in 2004, the TDR bank sold 9,842 square feet of Major Performing Arts Facility TDRs that generated $150,000 in debt payment for Benaroya Hall, 90,728 square feet of open space TDR putting $1.3 million toward the Seattle Art Museum's Olympic Sculpture Park, and 132,500 of Housing TDR directed at the preservation of downtown affordable housing (Seattle 2004).

References

Fesler, Stephen. 2018. Seattle Considers Changes to the City Incentive Zoning Program. The Urbanist. June 7, 2018. Accessed 3-8-21 at Seattle Considers Changes to the City Incentive Zoning Program | The Urbanist.

Kinney, Jen. 2016. Seattle Preserves Farmland, Funds Infrastructure by Building Taller Condos. Accessed 3-12-21 at Seattle Preserves Farmland, Funds Infrastructure by Building Taller Condos – Next City.

Massachusetts. Undated. Smart Growth/Smart Energy Toolkit – Transfer of Development Rights (TDR) Case Study: Seattle, WA.

Meier, Dennis. 2010. Correspondence with author of July 5, 2010.

Seattle. 2004. City Gains Housing, Debt Funding Through Sale of Transferable Development Rights. Seattle City Council News Release dated 4/19/2004.

Walker, Laura Hewitt. 1998. Correspondence with author November 25, 1998.

Shrewsbury Township, York County, Pennsylvania, population 6,715 (2018), uses TDR to preserve farmland through transfers to zones that increase density from six to eight units per acre on receiving sites with public sewer and water.

Smithtown, New York, population 116,384 (2018), motivates the preservation of open space, environmentally sensitive areas, and groundwater resources by allowing owners of sending sites to transfer unused water flow rights to receiving sites on the Transfer of Density Flow Rights Map or to the town's TDFR Bank.

Snohomish County, Washington, population 822,083 (2019), uses TDR to preserve farmland, forests, open space and natural resources. In sending areas, TDR allocation differs depending on the land use designations; for example, farmland is allocated more TDRs per acre than forest land. The TDR certificates identify the land use designation of their sending sites and, in receiving areas, the amount of bonus development is higher for farmland TDRs than non-farmland TDRs. As of October, 2020, 44 acres of farmland had been preserved by this program and the resulting 14 TDRs had been purchased by developers.

Snohomish, Washington, population 9,976 (2019), uses TDR to preserve resource lands, critical areas and open space. The receiving area is the Pilchuck District Center Zone where a three-story building height baseline can be increased to four or five stories at the ratio of 14,000 square feet of bonus floor area per TDR.

South Burlington, Vermont, population 19,162 (2019) uses a planned unit development process within its Southeast Quadrant to preserve open space, natural resources, scenic views, and agricultural resources and achieve up to eight units per acre on receiving sites. As of 2020, 114 units had transferred.

South Lake Tahoe, California, population 21,939 (2019), occupies the southern end of Lake Tahoe at an elevation of 6,237 feet in the Sierra Nevada Mountains. South Lake Tahoe is within the watershed of Lake Tahoe, which is internationally recognized for its outstanding clarity. In order to restore and protect water quality and aquatic ecosystems, development in South Lake Tahoe and five counties is regulated by the Tahoe Regional Planning Agency (TRPA) and its TDR program. This TDR mechanism is described in the profile of the TRPA program. South Lake Tahoe is given this separate profile because of the city's success at transferring TDRs from environmentally sensitive sending sites to urban receiving sites.

Like other locations in the Lake Tahoe watershed, South Lake Tahoe experienced a building boom over half a century ago, much of which was spurred by changes in gambling regulations and the 1960 Winter Olympics held in Squaw Valley, a few miles north of the city. Much of this development consisted of motels and hotels built prior to the imposition of modern regulations for storm water management. These older tourist units were also discouraging the development of new lodging and depressing the city's potential for greater tax revenues considering that the region is a winter sports mecca and the city itself is home to Heavenly Mountain Ski Resort.

The South Lake Tahoe Redevelopment Agency was at one time a major reason for the city's successful TDR program. The Agency acquired many older, nonconforming properties. Some of these acquired properties were recycled for new development using modern environmental standards. In some instances, an acquired site was unsuitable for development and was restored to its natural state in order to promote water protection benefits that in turn protect the water quality of Lake Tahoe. Importantly, the Agency banked marketable development rights from the sites it acquired.

The entire South Lake Tahoe redevelopment effort carried a price tag of $230 million. Much of this cost was recouped by improved property tax receipts and a transient occupancy tax of 12.05 percent on the room rates of tourist units. Importantly, the cost of acquiring the marketable rights needed to build new tourist units was offset by sales of the rights that the Agency acquired, such as the sale of 400 tourist unit rights and 6,000 square feet of commercial floor area rights to a single Embassy Suites hotel complex for $3.8 million (Solimar 2003).

The Redevelopment Agency also facilitated transactions by serving as a TDR bank where developers knew they could acquire marketable rights without having to find willing sellers and go through the time and expense of securing the necessary approvals. Furthermore, the Agency was able to write down the cost of these marketable rights in other to make redevelopment projects financially viable (Solimar 2003).

By 2001, the Redevelopment Agency had acquired 781 Tourist Accommodation Units (TAUs) and used 696 TAUs in downtown revitalization projects. Many of these TAUs resulted from the acquisition of environmentally sensitive sites that had been inappropriately developed under pre-TRPA regulations. Some of these sites underwent a natural restoration process to provide habitat and/or water protection features including storm-water retention basins. Alternatively, some properties in non-sensitive locations were reused for downtown revitalization but using development techniques designed to safeguard the clarity of Lake Tahoe (Solimar 2003).

Reference
Solimar Research Group. 2003. Tahoe Basin Marketable Rights Transfer Program

South Middleton Township, Cumberland County, Pennsylvania, population 15,352 (2016), uses TDR to preserve farmland, woodlands, riparian areas with associated wetlands and floodplains, plus historic/cultural/scenic resources.

Southampton Town, Suffolk County, New York, population 58,314 (2018), includes TDR provisions under various sections of its zoning code aimed at lowering or eliminating development in groundwater recharge zones, prime farmland, wetlands, areas designated for parks, greenbelts, and other public recreation, areas in the Old Filed Map Overlay District, and land within the core preservation area of the Long Island Central Pine Barrens Overlay District. The town adopted a Community Preservation Project Plan that allows revenue from a two-percent real estate transfer tax to fund a TDR Clearinghouse and Bank

Southold Town, Suffolk County, New York, population 22,125 (2018), preserves farmland, natural areas, open space, recreational land, and rural/historic/cultural landscapes by encouraging sending area property owners to record permanent conservation easements on sending sites and transfer sanitary flow credits from the preserved property to the town's flow credit bank. A sanitary flow credit is equivalent to the right to develop a single-family residential parcel with an individual on-site sewerage system or its non-residential wastewater equivalent. The bank sells these credits exclusively to add density to affordable housing receiving site projects. As of December 2019, the bank had acquired 65.99 flow credits and sold 10 in 2006 for a total of $125,000.

Springfield Town, Dane County, Wisconsin, population 2,900 (2020), has a TDR program offering higher transfer ratios to super sending areas and super receiving areas that receive high scores for achieving program goals.

Springfield Township, York County, Pennsylvania, population 5,792 (2018), uses TDR to preserve farms, prime agricultural soils, and natural resources with an amended ordinance calling for the establishment of a TDR bank and registry.

Stafford County, Virginia, population 152,882 (2019), uses TDR to preserve agriculture, forestry, rural open space, natural resources and scenic views. In one of four receiving zones, TDRs can boost density by over twelve units per acre or increase non-residential floor area by 3,000 square feet per TDR. TDR can be used as a matter of right. The program had saved 132 acres as of 2018.

St. Lucie County, Florida, population 328,297 (2019), uses TDR to preserve agriculture, habitat, flow ways, parks, recreation, and civic spaces. A dozen transfer ratios vary based on sending site priority and receiving site location.

St Mary's County, Maryland, population 113,510 (2019), forms the tip of land where the Potomac River meets the Chesapeake Bay in southern Maryland. There are approximately 43,000 acres of prime soils scattered throughout the County, with the largest tracts of prime farmland in the southern part of the County. The County has remained primarily rural despite the fact that it lies only 40 miles southeast of Washington, DC

St. Mary's County included a traditional transfer of development rights section in its zoning ordinance in 1992 in order to provide flexibility and preserve farmland and resource protection areas. These provisions preserved roughly 670 acres in the next ten years. In 2002, a TDR code amendment changed base zoning regulations, which increased activity, preserving more than 1,000 additional acres of land in the next four years, primarily on small, grandfathered lots that were severely constrained by environmental restrictions (St. Mary's County 2017).

Another amendment in 2007 reduced ways of circumventing the use of TDRs. For example, the county eliminated bonus density for design standards, LEED certification, or extra green space. As of 2007, developers could exceed receiving site base density using either actual TDRs or cash in-lieu. As of 2011, the cash-in-lieu option had generated $234,000. In 2016, the Maryland Department of Planning recommended that the transfer formula be simplified. However, the Comprehensive Zoning Ordinance on the County's website in 2021 remains unchanged with regard to Chapter 26 TDRs (Maryland 2016; St. Mary's County 2017).

Sending areas are parcels in Rural Preservation Districts (RPDs), a zone that allows only one dwelling unit per parcel, regardless of size, unless the site becomes a TDR receiving site as explained below. One TDR is allocated for every five acres of RPD-zoned gross land area. Sending site owners who choose to participate can choose to build on site or remove some or all of their development rights and transfer them to any person or legal entity.

TDRs can be used to increase residential density on parcels zoned RPD, RL, RH, RMX, VMX, TMX, CMX and RNC (in growth areas only.) The baseline density in these districts is one unit per acre with the exception of 10 units per acre in the RH and special requirements that apply in the RPD (explained below.) In all but the RPD, one TDR increases residential density by one dwelling unit. The increased density allowed through TDR ranges from 1 bonus unit per acre in the RNC district to

14 bonus units per acre in the CMX district. On receiving sites zoned RPD, the first dwelling unit built on a receiving site does not require a TDR but uses five acres of base density. Thereafter less acreage is needed for each bonus lot but each bonus lot requires more TDRs as depicted in the following table.

Density (units per acre)	Number of TDRs
1 dwelling unit per 5 acres	1 TDR per lot after the first lot or dwelling
1 dwelling unit per 4 acres	2 TDRs per lot or dwelling
1 dwelling unit per 3 acres	3 TDRs per lot or dwelling

TDRs can be also be used to increase non-residential intensity of land zoned RPD, RSC, RCL, RL, RMX, VMX, TMX, DMX, CMX CC, OBP and I at the ratio of 2,000 square feet of floor area in excess of a baseline FAR for each TDR.

Categories of receiving areas are paired with categories of sending areas. For example, receiving areas can only be located in a Critical Area if they use TDRs from sending areas in the Critical Area. Similarly, TDRs cannot be transferred to a Resource Conservation Area from within an Intensely Developed Area (IDA) or Limited Development Area (LDA) or from an IDA to an LDA.

St. Mary's County allows developers to comply with TDR requirements using cash in lieu of TDRs. The amount is set at 120 percent of the average fair market value paid for TDRs in private market transactions in the previous year, but the Board of County Commissioners can increase or decrease that amount. In 2011, the Board set the cash-in-lieu amount at $20,000. The County can use the cash-in-lieu payments to buy development rights or replenish the Critical Farms Program.

By 2016, St. Mary's County's TDR program had preserved 4,107 acres, mostly with transfers between rural sending and receiving sites (Maryland 2016).

References

Maryland. 2016. Transfer of Development Rights Committee Report. Accessed 2-3-21 at TDR-committee-report-2016.pdf (maryland.gov).

St. Mary's County. 2017. Land Preservation, Parks, and Recreation Plan Update. Accessed 2-3-21 at Approved 2017 LPPRP.pdf (stmarysmd.com).

St. Petersburg, Florida, population 261,338 (2019), has an environmental protection TDR program that motivates transfers from developmentally-constrained Preservation Areas. The city also has a historic preservation TDR program allowing transfers from designated local landmarks.

Stearns County, Minnesota, population 161,075 (2019), uses TDR to protect farmland and natural resources. Proposed sending or receiving sites must be approved by the applicable township. TDRs cannot be transferred to intermediaries and a TDR bank is not allowed.

Stowe, Vermont, population 4,314 (2010), uses TDR to preserve resource lands and rural character by transferring TDRs that can increase building coverage, residential density or lodging unit density in seven receiving areas.

Suffolk County, New York, population 1,477,000 (2019), manages three TDR programs that preserve land for open space and groundwater quality protection by sanitizing sending sites and creating credits that are transferred to receiving sites to build workforce housing and/or to exceed baseline densities in areas often not served by sanitary sewers. Each credit represents 300 gallons per day of sanitary flow which allows one single-family detached residential unit, two attached units with 600 square feet of floor area or less, 1.3 attached units less than 1,200 square feet, three Planned Retirement units less than 600 square feet, two retirement units greater than 600 square feet, and amounts of floor area for various non-residential uses. By 2014, these programs had created 658 credits. Suffolk County is also home to the Central Pine Barrens TDR Program and separate TDR programs adopted by seven different towns within Suffolk County.

Summit County, Colorado, population 31,011 (2019), surrounds the popular Breckenridge ski resort in central Colorado. Summit County adopted four TDR programs to protect environmental resources in the county's four watershed basins. The Upper Blue Basin TDR Program is the most active and had protected 2,000 acres as of March 2021.

Summit County's Climate Action Plan, adopted in 2019, recognizes that forest preservation and maintenance protect existing carbon sinks and promote the sequestration of future carbon emissions. Consequently, the plan calls for expansion of the existing TDR programs in the Lower Blue, Snake and Ten Mile river basins and encouragement of the towns of Dillon, Frisco, and Silverthorne to use TDRs to protect forested lands from development.

The Upper Blue Basin TDR Program, launched in 2000, operates within the Upper Blue Basin which encompasses approximately 80,400 acres and has a permanent population of roughly 9,500 inhabitants. Development here mostly occurs in the valley near the Blue River, typically near the towns of Breckenridge and Blue River. Almost 80 percent of the basin lies within the National Forest Service system, primarily consisting of forested mountainsides. However, the backcountry is also dotted with hundreds of private mining claims that are technically capable of residential development even though they are often located on ridgelines and above the timberline in environmentally sensitive areas. Development here can also degrade important viewsheds and block access to national forest recreational sites. Consequently, the TDR program initially aimed to protect backcountry areas, natural resources, and open space in the mountains near Breckenridge and Blue River.

The county attributes the success of the Upper Blue Basin TDR Program partly to the participation of the towns of Breckenridge and Blue River. Importantly, town and county policies prohibit upzonings without the use of TDR. In addition, owners of private mining claims are motivated to participate in transfers by the adoption of the Backcountry Zoning District, which rezoned hundreds of mining claims to a maximum density of one dwelling unit per 20 acres. Each TDR here represents the preservation of 20 acres of land.

In the Backcountry zoning district, existing roads cannot be improved without a conditional use permit and new roads must comply with standards that minimize disturbance. Snow plowing in this district also requires a conditional use permit and can only occur if at least four inches of snow is left on road surfaces. The county may not provide public facilities, emergency response, or communities services to properties in this zone.

Single family dwellings in the Backcountry zone are limited to 2,400 square feet of floor area. Dwellings on legal nonconforming lots of two acres or less are limited to 750 square feet of floor area and can gain an additional 50 square feet for each acre in excess of two acres up to a maximum of

2,400 square feet. In addition, extensive requirements apply to accessory structures, decks, tree removal, and other site alterations.

The county designates sending and receiving areas on the Official Transfer of Development Rights Map. According to Code Section 3506.02, TDR Regulations, owners of land within the mapped receiving areas can apply to use TDR in conjunction with a zoning amendment or PUD modification that would increase density, increase floor area, increase vehicle trips for commercial/industrial uses, and/or increase the activity levels allowed by the applicable zone. Applicants can also propose TDR as a means of mitigating development impacts. Various types of development are exempt from these provisions including affordable housing and modifications of internal density within the Keystone and Copper Mountain Resort PUDs.

Each TDR allows receiving site baselines to be exceeded by one single-family dwelling unit of up to 4,356 square feet, or one multi-family unit of up to 1,400 square feet, or three rooms of lodging not to exceed 467 square feet each, or 1,000 square feet of non-residential floor area. When receiving sites are proposed to exceed vehicle trip baselines, each TDR allows the number of trips generated by a single-family residence as stated in the most recent edition of the Trip Generation Manual of the Institute of Transportation Engineers.

Summit County prefers transfers between sending and receiving sites in the same basin. However, the review authority for the receiving site basin may consider an inter-basin transfer based on five criteria including conformance with planning goals and evidence that the needed TDRs are not readily available within the same basin. The code section on inter-basin transfers adds further conditions about where, when and how many TDRs can be transferred between specific basins.

In the Lower Blue, Snake River, and Ten Mile basins, developers may voluntarily propose alternatives to the TDR program requirements such as dedication of land for open space, community facilities, or affordable housing. Such proposals are subject to five criteria including that the value of the proposed community benefit be roughly proportional to the value of the number of TDRs that the project would otherwise have to acquire. In addition, the county allows contributions in lieu of TDRs that are used to buy development rights from sending area properties.

The TDR Bank is also a major success factor. The Bank has a separate account for each of the four TDR programs operating in Summit County's four basins. The Bank's sale price for a TDR from the Lower Blue River Basin account is the fair market value of that TDR determined on a case-by-case basis. The Bank sells Upper Blue Basin TDRs at prices established annually based on the median sales price of all vacant land zoned Backcountry in that basin sold after 2000, the year when that TDR program was instituted. The Bank sells TDRs from the Snake River Basin and Ten Mile Basin at prices set each year and based on the median sales price from all vacant land zoned Backcountry since 2007, the year that the county adopted Backcountry zoning in those two basins. In March 2021, the bank sold Upper Blue TDRs for $99,045, roughly $4,952 per acre and Countywide TDRs for $63,065, or roughly $3,153 per acre.

The Upper Blue Basin TDR Program had preserved 2,000 acres as of March 2021 and generated roughly $4 million for further open space protection. Program amendments now allow TDRs to also be sold in return for the preservation of high-quality wetlands in receiving areas. As of March 2021, the program had protected roughly 14 acres.

In addition to the 2,000 acres protected by the Upper Blue Basin TDR Program, as of March 2021: the Snake River Basin TDR Program, adopted in 1998, had protected roughly 300 acres in nine separate transactions; the Ten Mile Basin TDR Program, adopted in 2006, had protected 199 acres in a single transaction; and the Lower Blue Basin TDR Program, adopted in 2007, had protected 20 acres in one transaction.

Summit County, Utah, population 42,145 (2019), includes a Land Bank and Development Right Relocation process in the Snyderville Basin Development Code granting variable bonus density based on the public benefit gained by a transfer. It saved over 1,000 acres partly by traditional TDR and partly by preservation funded by the sale of county land acquired by transfer.

Sunderland, Massachusetts, population 3,659 (2018), allows 1:1 transfers from permanently preserved agricultural and watershed protection sending sites to receiving areas where TDR can increase density by a factor of no more than two.

Sunny Isles Beach, Florida, population 21,942 (2019), uses TDR to redevelop blighted areas, fund infrastructure, and create educational/recreational facilities including parks and open space. In addition to privately-owned sending sites, the city can sever and sell TDRs from city owned parks established after incorporation. The city bought non-waterfront land for parks and sold the resulting TDRs for extremely profitable prices to the developers of waterfront receiving area projects who found TDRs to be more affordable than buying land. As of 2009. the program had generated $35 million, which funded nine parks, various recreational programs, a school, and several public works projects.

Sussex County, Delaware, population 234,225 (2019), adopted a density bonus fee to fund open space preservation and park land. In 2020, the fee was $15,000 per unit above two dwelling units per acre in 14 town centers and developing areas or $20,000 per unit in excess of two units per acre in coastal areas.

Tacoma, Washington, population 217,827 (2019), uses TDR to achieve three distinct benefits. Sending areas that protect habitat within the city can sever one TDR for each foregone dwelling unit which can be used to gain 15,000 square feet of bonus floor area on receiving sites. In the landmarks TDR program, one TDR is issued for each 600 square feet of allowed but unused floor area. In the interjurisdictional program, each TDR transferred from sending sites in unincorporated Pierce County allows 5,000 square feet of bonus floor area and each TDR from King County allows 10,000 square feet of bonus floor area. Developers can also receive one square foot of bonus floor area for every $2 paid to the city's open space fund, which must be used to buy TDRs from city or regional sending sites. In 2016, Tacoma celebrated its first interjurisdictional transfer which helped preserve a Pierce County farm.

Tahoe Regional Planning Agency (TRPA) regulates land use and environmental protection for a region that encompasses the City of South Lake Tahoe, California, plus two counties in California and three counties in Nevada. The TRPA is tasked with improving and protecting the water quality of Lake Tahoe, which still retains extraordinary clarity despite attracting millions of visitors, including winter sports enthusiasts who flock to the ski resorts surrounding the lake. The region consists of the entire 207,000-acre Tahoe Basin which TRPA manages using various initiatives and regulations including a complex TDR program.

Lake Tahoe's clarity has been recognized by designation as an outstanding national resource under the federal Clean Water Act. However, the lake and region were threatened by development spurred by the rising popularity of skiing, the 1960 Winter Olympics held here at Squaw Valley Ski Resort, and federal gambling regulations that made the Nevada side of the basin extremely attractive for large casinos. California and Nevada recognized that the environment could not withstand massive growth and signed a bi-state compact which was ratified by the US Congress in 1969. TRPA was subsequently launched and ultimately adopted regulations for land coverage and growth rates that incorporated programs for transferring rights. These programs, as described immediately below, originally limited conversion between various land uses (residential, tourist lodging, commercial) and contained approval requirements for transferring rights between jurisdictions. However, subsequent revisions have relaxed or eliminated some of the limitations originally placed on the region-wide TDR market, as discussed at the end of this profile.

Land Coverage Transfers – In 1987, TRPA promulgated a set of ordinances that addressed allowable land uses, rates of development, density, scenic impact, and land coverage limits aimed at minimizing water runoff, removing contaminants, and reducing erosion on land with a wide range of permeability due to soil type and slope. In some areas, up to 30 percent coverage might be allowable without creating degrading amounts of sediment runoff and erosion. Conversely, the most sensitive sites in the Stream Environment Zone (SEZ) might be confined to as little as one percent of a site's total land area. These coverage regulations can constrain the ability to build new structures or expand existing buildings.

As mitigation, TRPA developed a land coverage transfer program offering property owners the option of buying coverage rights from sending sites that permanently preclude excess coverage. The sending sites must be classified as more sensitive than the receiving sites and both sending and receiving sites must be located in the same hydrologic zone. There are nine hydrologic zones in the basin. Generally, the transfer ratio is one-to-one, meaning the amount of bonus coverage allowed at the receiving site is equal to the amount of coverage prohibited at the sending site. The maximum coverage allowed with transfers varies depending on whether the receiving site project is commercial, tourist lodging, public facilities, or residential development (which also can vary depending on the size of the project). Receiving site owners can buy land coverage rights on the private market or from the California Tahoe Conservancy, which manages a land coverage bank.

Transfers of Allocation – In order to keep growth from overwhelming the capacity of public infrastructure and services, TRPA sets annual limits on the amount of development allowed in the basin. For example, the quota might be 300 dwelling units, 400,000 square feet of commercial development, and 200 rooms of tourist lodging units. In this transfer mechanism, the sending site must be vacant, assigned a land capability classification that is so sensitive the site is ineligible for development, and permanently precluded from development by either deed restriction or ownership by a public or private non-profit agency tasked with open space preservation. The receiving site must have a less sensitive land capability rating than the sending site and be planned for residential development,

Transfers of Development Rights – TRPA requires each new residential unit to have a development right as well as an allocation. To facilitate these acquisitions, TRPA allows residential development rights to be transferred within various parameters. TRPA aims to remove inappropriate existing development from sensitive sending sites by allowing owners to sell development rights for removing existing structures and returning the sites to a reasonably natural state. This removal/restoration process creates both a development right and an allocation because the transfer creates no net increase in development. Under TRPA's quota system, these transfers have the potential to generate a strong incentive to remove inappropriate development from sensitive areas.

Recent Amendments – Until 2018, there was limited opportunity to convert rights between three land uses: residential units of use (RUU), commercial floor area (CFA), and tourist accommodation units (TAU). In 2016, TRPA hosted a Development Rights Strategic Initiative aimed at addressing low convertibility and other constraints on the development rights marketplace that were slowing down the replacement of older, environmentally harmful development with new construction designed to minimize environmental impacts. Five main recommendations evolved from the initiative.

- Use environmentally neutral exchange rates to allow conversions between different development rights: RUU, CFA and TAU.
- Expand opportunities to qualify for the residential bonus unit incentive program.
- Expand the development rights banking system.
- Remove overlapping, multijurisdictional approval requirements for development rights transfers.
- Remove the requirement to have an approved project on a receiving site prior to the transfer of development rights.

In 2018, the TRPA Governing Board adopted these recommendations.

- Section 51.4.3 of TRPA's Code of Ordinances allows for the conversion of CFA, TAU, Single Family Existing Residential Units of Use (SF ERUU), and Multi-Family Existing Residential Units of Use (MF ERUU) using a table of 16 possible conversion ratios. For example, 300 square feet of CFA from a sending site can be used at a receiving site to add 300 square feet of CFA, one TAU, 1 SF RUU, or 3/2 MF ERUU.

- The Code allows a portion of the residential bonus pool to be allocated to units of "achievable housing", meaning households with above median income yet not able to afford median priced housing. In addition, Section 51.5.1.C.3, Transfer of Potential Residential Units of Use to Centers; Bonus Unit Incentive, grants additional units to receiving areas in centers according to a formula that increases the transfer ratio based on the land capability district of the sending site and the distance of the sending site from centers and primary transit routes. Chapter 52 additionally uses a points system to award bonus density to receiving sites that participate in various mitigation measures such as environmental improvement programs for transportation, water quality, and SEZ restoration as well as access to outdoor recreation sites and scenic quality improvement programs approved by TRPA.

- The 2018 amendments eliminated the requirement for approval from local government of both the sending and receiving sites, a requirement that was recognized as complex and costly. However, a local government can ask the TPRA to create a local approval process if the net loss of development rights resulting from transfers over a two-year period is equal to or greater than five percent of the total existing built development rights for each type of land use (CFA, TAU, RUU) within that jurisdiction.

- TRPA committed to partnering with the California Tahoe Conservancy, the Nevada Division of State Lands, and private non-profit conservancies to increase the development rights holdings in TDR banks.

- The 2018 amendment increased flexibility and user-friendliness by eliminating the need to have an approved project prior to a transfer. TRPA also committed to expanding its development rights tracking and inventory capabilities and highlighting redevelopment success stories.

Tarpon Springs, Florida, population 25,176 (2019), allows one transferable dwelling unit per acre or 0.5 FAR per acre to be transferred from conserved wetlands and used to achieve maximum future land use map density/intensity on receiving sites. In addition, a Smart Code motivates the preservation of historic resources and the fishing/shrimping industry in the Sponge Docks redevelopment district by allowing density transfers to receiving sites in eleven transect zones.

Teton County, Wyoming, population 23,464 (2019), uses non-contiguous planned residential development to preserve open space by transferring density from sending to receiving parcels. Used at least once to preserve 237 acres.

Thurston County, Washington, population 290,536 (2019), uses TDR to preserve farmland with intra-jurisdictional transfers and interjurisdictional options in the cities of Olympia, Tumwater, and Lacey. Olympia allows TDRs to be used to either increase density from seven units to eight units per acre or decrease density from five units to four units per acre in its R 4-8 Zoning District. Transfers that increased density preserved 181 acres with TDRs averaging $17,000 each.

Townsend, Massachusetts, population 9,547 (2018), uses TDR to preserve water resources and open space with density transfers to two receiving districts.

Troy (Town), St. Croix County, Wisconsin, population 4,757 (2012), adopted a TDR program when its citizens balked at paying additional taxes to fund a purchase of development rights program for agricultural preservation.

Vancouver, Washington, population 184,463 (2019), uses TDR to preserve historic landmarks by transferring unused development potential to receiving sites in the same zoning district. The only limit on with-TDR development is that receiving site projects not create a hazard to low-flying aircraft.

Wareham, Massachusetts, population 22,666 (2018), uses TDR to preserve open space, natural resources and historical features. By special permit, TDRs allow receiving sites to use modified requirements for lot size, frontage, lot depth, setbacks, lot coverage and building height as well as achieve bonus density.

Warrington Township, Bucks County, Pennsylvania, population 24,474 (2018), has successfully used TDR since 1985 to preserve open space, farmland, environmentally-sensitive areas and historically-significant sites.

Warwick (Town), Orange County, New York, population 31,185 (2018), preserves farmland with traditional TDR and Incentive Zoning which lets developers pay $50,000 per bonus dwelling unit on a receiving site that the town uses to purchase land or development rights in sending areas. The Town of Warwick and Warwick Village have an annexation policy agreement establishing how revenues from cash payments in lieu of open space will be divided: 25 percent for areas in the town that protect the village watershed, 45 percent decided solely by the village, and 30 percent under the sole control of the town.

Warwick Township, Lancaster County, Pennsylvania, population 19,323 (2019), is located 75 miles west of downtown Philadelphia in Lancaster County. For decades, Lancaster County has been a national pacesetter in farmland preservation and Warwick Township leads the county, with 3,060 acres permanently preserved by several tools including TDR.

Warwick adopted a TDR ordinance in 1993 which was subsequently amended in 1997, 2001, 2009, and 2016. The program aims to protect the township's agricultural economy and landscape as well as farmland itself.

Owners of land in the Agricultural Zone who choose to participate are issued one TDR per two acres of qualifying land preserved by permanent easement. Sending site owners can sever and sell some or all of their TDRs. When transferring only some of their TDRs, the sending site owner must submit a plan identifying the portion of the site where TDRs have been severed, the number of TDRs severed and the number remaining on the site.

As of 2021, the program offers three types of incentives to developers of receiving sites. In the Campus Industrial Zone, TDR developments can exceed a baseline of 10 percent lot coverage and achieve a maximum lot coverage of 70 percent by using one TDR for each 4,000 square feet of additional lot coverage. Buildings in this zone can also exceed a baseline height of 45 feet and attain a maximum height of 65 feet by using one TDR for each 4,000 square feet of additional lot coverage that would have been needed if the building were limited to 45 feet in height.

In a third receiving site mechanism, housing-for-older-persons developments in the R-3 Residential zone can exceed a baseline of five units per acre and achieve a maximum density of 14 units per acre using one TDR per bonus unit per acre.

As expressly stated in the zoning code, Warwick can buy TDRs and accept TDRs as gifts. The township formed a bank initially capitalized with general fund money which has been replenished by income from the sale of its TDR holdings. The code also allows private, non-profit conservancies and the Lancaster County Agricultural Preserve Board to buy and resell TDRs as long as the sale proceeds are used exclusively to buy TDRs from Warwick sending sites. In effect, this process transforms what is typically a purchase of development rights (PDR) process into a TDR process. PDR programs simply retire development rights, making it necessary for more public money to be raised before additional rights can be purchased. Conversely, in Warwick Township, money from the Lancaster County Ag Preserve Program can be pooled with funds from the Lancaster Farmland Trust and Warwick itself to buy TDRs that can then be resold, turning what would otherwise be a single acquisition into a perpetual farmland preservation revolving fund.

As of 2021, Warwick had preserved 1,617 acres, or over half of its 3,060-acre preserved farmland total, using TDR. The 3,060 acres of preserved land represents 44 percent of the total land area in

Warwick's Agricultural Zone. The township TDR bank alone purchased 668 TDRs and sold 450 TDRs as of 2021. A total of 833 TDRs have been purchased and 592 TDRs sold through Warwick's cooperation mechanism with the Lancaster Farmland Trust and the Lancaster County Agricultural Land Preservation Board.

Waseca County, Minnesota, population 18,740 (2019), preserves farmland with a TDR program that allows transfers without CUP between properties in the same township and by CUP between properties under separate ownership in different townships. Activity has occurred and a sunset provision was eliminated.

Washington, DC, population 692,683 (2019), adopted a plan in 1984 aimed at transforming the downtown into a vibrant, mixed-use, 24/7 neighborhood where people could live, shop, experience art, and enjoy historic landmarks as well as work. To implement those goals, the city organized the downtown into various districts for landmarks, art, retail, housing, and Chinatown. Beginning in 1989, developments in the retail overlay received bonus density for dedicated retail space. In 1991, the bonus development provisions were expanded to the other districts and to historic structures. The height limitations imposed by Washington's Height Limit Act of 1910 make it difficult or impossible to use the bonus density on site. Consequently, the city allows developers to transfer density to off-site receiving areas in the downtown and areas on the outskirts of downtown.

Downtown Shopping District – Above a mandatory FAR 2.0 of retail, additional department store floor area generates FAR 3.0 of bonus density for the building. Bonus FAR 2.0 is generated for a legitimate theater. Bonus FAR 1.0 is created for floor area dedicated to a movie theater, performing arts space, anchor store, or minority-owned business.

Downtown Arts District – Here, FAR 1.0 must be devoted to entertainment, art, or retail, including at least 0.25 FAR in true art. Various FAR bonuses are granted for additional floor area in several qualifying art categories including art centers and art schools.

Chinatown – Bonus FAR is granted for specified uses as well as retail floor area above FAR 1.0.

Residential and Mixed Use Districts – Bonus density generated by various retail uses including groceries and drug stores.

Historic District – Unused development potential can be transferred to receiving sites.

Developers can provide the preferred uses that qualify for FAR bonuses by using a procedure called Combined Lot Development on two separate sites as long as they are located in the same district.

Developments are limited to 6.5 FAR and 90 feet of building height in the downtown. Consequently, TDRs can be transferred to two areas at the edge of downtown. In Downtown East, developments using TDR can achieve 9.0 FAR and 110 feet of building height. In the other receiving area, New Downtown, projects using TDR can reach FAR 10.0 and 130 feet of building height in places where that height is allowed under the 1910 Height Act.

Washington facilitates transfers by issuing TDRs upon the execution of covenants that incorporate a schedule for required renovations and maintenance of the preferred use as well as sending site restrictions. These TDRs can be held or applied immediately at a receiving site using an administrative approval process.

Between 1990 and 2007, 9,523,000 TDRs were generated. Of this total, 7,985,000 TDRs were transferred to receiving sites. Historic landmarks generated many of these TDRs until 2005. Beginning in 1998, residential TDRs began to predominate. In the early 1990s, TDRs averaged over $35 per square foot but dropped to about $10 per square foot in the early 2000s (Alpert 2009).

Reference
Alpert, David. 2009. Downtown's zoning: How the best of downtown came to be. Greater Greater Washington. Accessed 6-22-21 at   https://ggwash.org/view/1793/downtowns-zoning-how-the-best-of-downtown-came-to-be.

Waukesha County, Wisconsin, population 404,198 (2019), maintains a TDR mechanism in its zoning code that can be used by townships wanting to use this tool for the preservation of prime farmland and environmentally sensitive areas.

Weber County, Utah, population 260,213 (2019), has transfer components in its Ogden Valley Destination and Recreation Resort Zone, which is designed to protect specific shorelines, significant habitat, rural character, resources and farmland when approving resorts of at least 1,000 acres.

West Hempfield Township, Lancaster County, Pennsylvania, population 16,555 (2018), had preserved 345 acres of farmland as of 2008 using TDR.

West Hollywood, California, population 36,475 (2019), uses TDR to preserve and restore designated cultural resources. Sending site owners must escrow 75 percent of TDR sales proceeds to finance restoration work if needed.

West Lampeter Township, Lancaster County, Pennsylvania, population 15,944 (2018) uses TDR to preserve farms and environmental areas. In 2007, TDRs were donated to the Lancaster Farmland Trust which pledged to use the revenue from the sale of those TDRs for further preservation in West Lampeter.

West Palm Beach, Florida, population 109,767 (2019), uses TDR in its downtown to preserve historic landmarks and parks. As of 2010, TDR here had protected as many as 15 properties including two churches.

West Pikeland, Chester County, Pennsylvania, population 4,079 (2018), uses TDR to preserve farmland, natural areas and community character. On receiving sites, one TDR yields a bonus of 1.1 single family detached units, 1.4 attached units, 1.7 multiple family units or two mobile home units.

West Vincent Township, Chester County, Pennsylvania, population 5,726 (2018), uses TDR to preserve farms, environmental areas, open space, views, and historic character. Bonus development ranges from one unit per TDR to 2.5 units per TDR depending on the zoning of the sending and receiving sites. In non-residential zones, one TDR yields 5,000 square feet of additional floor area.

West Windsor Township, New Jersey, population 28,045 (2018), adopted TDR to permanently preserve a private golf course for recreational open space.

Westborough, Massachusetts, population 18,272 (2010), uses TDR to preserve natural features and open space. Receiving sites in the Transit-Oriented Village district gain ten bonus units per preserved acre in the single family residential zone or five bonus units per preserved acre in other zones.

Westfield, Massachusetts, population 41,449 (2019) uses TDR to preserve farmland, water resources, and rural/historic character. Developers increase floor area, lot coverage, and building height on receiving sites in four non-residential zones by recording easements on sending sites or making a cash contribution based on the average cost of development restrictions over the last three years.

Westlake, Texas, population 983 (2019), adopted Transfer of Development Intensity (TDI) regulations in 2017 aimed at preserving view corridors, managing traffic, and implementing the open space and public facility needs of the comprehensive plan. All privately owned land in Westlake is mapped as sending, receiving, or dual eligibility districts. Only land zoned planned development can participate. Bonus development is granted for public art and donations of land to the public.

Whatcom County, Washington, population 229,247 (2019), uses TDR primarily to preserve the watershed of Lake Whatcom. A 2018 study reported that 247 TDRs had been certified but only 18 had transferred due to low demand for higher density, lack of relevant incentives, limited county receiving sites, interjurisdictional participation only from the City of Bellingham, the uncertainty/complexity of a discretionary approval process, an inactive cash-in-lieu option, and the additional development cost of TDR. The study urged expanding the density transfer charge option due to its simplicity and flexibility.

Wicomico County, Delaware, population 103,609 (2019), uses TDR to preserve farmland. Sending area allocations are higher for receiving sites within growth areas than allocations of TDRs used at receiving sites outside of growth areas.

Williston, Vermont, population 9,870 (2018), allows 1:1 transfer of development potential from conservation areas. Unlike some programs, TDR here cannot be used for deviations from the town's annual building permit quota system.

Winchester/Clark County, Kentucky, population 36,263 (2019), uses TDR to shift growth to appropriate places. Transfer ratios increase from 1.5 to 2.5 as development potential moves from quadrants least suitable to quadrants that are most suitable for additional growth. It has been used at least twice.

Windsor, Connecticut, population 29,069 (2014), uses Section 4.5.8. Transfer of Residential

Density to shift growth from sending sites with potential public benefit to six receiving zones where density can increase from one unit per acre to five units per acre. Preservation and development approvals are concurrent.

Woolwich Township, Gloucester County, New Jersey, population 12,786 (2018), uses TDR to preserve farmland and environmentally sensitive areas. In 2016, Woolwich's TDR bank used local and state funding to buy 274 credits by reverse auction which preserved 820 acres of farmland.

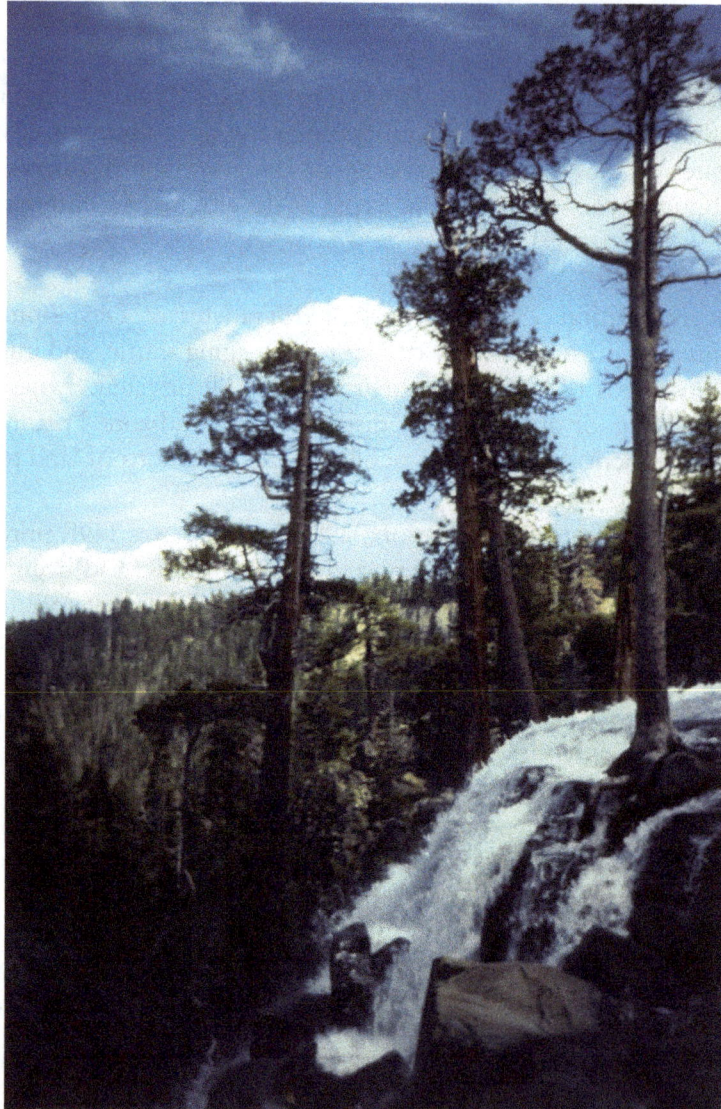

The Tahoe Regional Planning Agency uses various TDR mechanisms to help protect the clarity of Lake Tahoe.

ACKNOWLEDGEMENTS

I am grateful to the many planners and scholars, unfortunately too numerous to name, who have helped me understand TDR concepts and programs since I began to delve into this topic in 1978.

I am also indebted to my wife Adrian for her detailed proof reading and to Creative Juices for skillfully designing and producing this book.

INDEX